Cannon Hall Farm
COOKBOOK

Seasonal recipes and stories from a
Yorkshire farmhouse kitchen

Richard Nicholson

MB

MIRROR BOOKS

Published in Great Britain and Ireland in 2025 by
Mirror Books, a Reach PLC business,
5 St Paul's Square, Liverpool, L3 9SJ.
www.mirrorbooks.co.uk
@TheMirrorBooks

ISBN 9781917439466

Printed and bound in Great Britain by
Bell & Bain Ltd, Glasgow, Bell & Bain Ltd

This book was printed using
FSC approved materials.

Photographs by Clive Shalice
Styling by Victoria Gray
Illustrations: Joanna Lisowiec/Folio
Cover design: Julie Adams
Typesetting & design: Danny Lyle

This book is dedicated to Clare. It could never
have been completed without her love and support.

And to Mum and Dad. Without the decades of selflessness,
dedication and hard work you contributed to our family
farm, we could never have made it this far.

Contents

Our Family Food Journey

For me, food is not fuel, to be added to an engine. Food is generosity and warmth, fodder for the soul.

The best cookbooks are about more than food – they're about life and how to live it!

For those for whom food is evangelism, based around a fixed set of rules that should not be broken, I have nothing more than a sad smile. If that's what food is to you, then move onward on your journey. No need to trouble yourself here.

A love of being creative, through drawing, painting, writing and, most often, cooking, has got me through some sticky times. I've faced challenges throughout life with my mental health. The death of my wife, Maxine, in 2009 at just 46 years of age was certainly one of those moments. I've lived through times that were full of doubt and uncertainty, where I questioned what life meant and where I was headed. What I've discovered is that if you love making and creating, you have a pathway to salvation. A diagnosis of ADHD at 57 explained a lot of things that I'd struggled with all my life.

I've found that an interest in food can provide a route to self-expression; a path that can steer you away from negativity, towards creativity and a help to guide you through difficult times.

Chatting with my mum and dad about their own food journeys, I've realised just how much I didn't know. How much of their lives before I came along that I wasn't aware of. It's been lovely to spend more time with them and hear their stories. It's been another lesson, learned later in life. Spend time with those you love while you can, and make sure you keep asking questions. When they're gone you won't be able to mine that seam any more, and as every Barnsley miner knows, there's a lot of valuable coal to be found, sometimes, you might even find a diamond.

Food has been a huge part of mine and my family's journey. Food is about search and discovery. Food is reward for curiosity. Food is inspiration and enlightenment. It has the power to transport us back through the decades. Smells, tastes, and recipes can bring back people, places and experiences from the past, in vivid technicolour.

At Cannon Hall Farm, we believe that food can be therapy, that the simplest of pleasures can create the most memorable of times.

We all love sharing our family recipes, and writing this book has brought back memories of people who are no longer with us, but who have played a precious part in our journey. There's nothing quite so evocative as food, and all families that love good food will have a well-thumbed, and much-loved recipe book.

This is ours.

This book is a journey into our family folklore, with tales from the past, alongside recipes that we've grown to love. A celebration of the dishes Mum loved to cook when we were kids, through to the dishes inspired by her, and other friends and family, and on to those that I love to make today.

That's not to say there aren't healthy dishes here, or ones that can be adapted to be healthier, however we all love good, old-fashioned, comfort food, and at Cannon Hall Farm there will always be a time and a place for that.

I write this in the hope that you will embrace your own food journey; that you'll write down those family recipes before it's too late, and they're lost forever. That's what I've been doing, and I hope you'll enjoy this book of recipes that goes back through the years, right up to the current day. My hope is that you'll adapt them, and scribble down your own variations of them in the margins. Most of all, I hope this recipe book becomes food splattered, and dog eared, in the way that the best cookery books do. I hope that ragged pieces of paper, and hastily written notes nestle within its pages, bookmarks on your own food journey. I hope it becomes loved; in the way we have loved the family and friends who played their part in it.

This book is many things. A collection of stories, and recipes, a self-help project, a memoir, but most important of all, a family cookbook.

At some stage along the way, and for many reasons, I feel that I lost a part of me. This book, the journey of writing it and the love that went into it, has helped me to find my true self once more.

My Food Philosophy

Anyone who has watched my Farm Shop Facebook Live streams will know that I'm not someone who sticks hard and fast to a recipe, but will adapt to suit the ingredients available at the time. A recipe is just the starting point on the journey, not necessarily where you end up. It's very rare that I'll follow a recipe to the letter and rarer still that I'll make it the same way again, even if I like what I've cooked. Life is about good foundations, use of the appropriate building blocks and adding attractive embellishments. I'll use a recipe as a guide, and then adapt it, adding ingredients I like, and sometimes substituting or leaving out others that I don't.

My advice is to read as much as you can about food and the way it's evolved through different influences around the world. Experiment with the sort of food you eat and the way you combine ingredients. Look at fusion cooking, that embraces more than one food style, and how it has combined to create new trends. The world of food is an ever-changing, interesting tapestry of ideas and influences. If you embrace it, your life will be so much richer.

One of my more recent loves is finding out about the wild foods that can be foraged in the fields, hedgerows and alongside the stream here on the farm. Many of these ingredients have been forgotten and are only now being rediscovered and repurposed for the modern age, adding variety to both cutting edge restaurants and home kitchens.

Cooking isn't some sort of dark alchemy. You will have experiments that go wrong but, for the most part, you will start to pick up on the foods and flavours that work well together. You'll encounter more little triumphs, and the notes in your cookbooks will become more frequent. That's the moment you can really start to become creative in the kitchen.

My Favourite Kitchen Essentials

Sea Salt

If you use ordinary table salt, just changing to sea salt will improve your cooking considerably. Both contain around 40% sodium, but sea salt is usually free of anti-caking ingredients, has a cleaner flavour and contains minerals and trace elements. The size of the particles is less uniform, so when crumbled onto food it adds extra bursts of flavour when used to season a dish.

Pestle And Mortar

Always pick a good, heavy, stone pestle and mortar. So many of those sold are inadequate. Those made of wood or ceramic are not usually up to the job. You need real grinding power. I like to start off my salad dressings by grinding up sea salt and a small clove of raw garlic. This helps to release the oils from the garlic, intensifying the flavours to create a flavoursome base for the other ingredients.

Potato Ricer

A good potato ricer guarantees smooth mashed potato every time. Cook your potatoes, drain and allow to sit in the pan for a few minutes, then pass them through the ricer. Add butter, a dash of milk, salt and black pepper and beat until smooth. I often add spring onion, roasted garlic or wholegrain mustard to my mash to give variety. Or, sometimes, I'll gently fry leeks, onions or shallots in butter until soft and add those.

Dutch Oven

This large cast iron pot with a lid is super useful, adaptable and practical for both indoor and outdoor cooking. It's my first choice for stews and braising and you can even bake bread in it. You can use it on the stovetop to sear stewing cuts before moving to the oven. You can use it over an open fire outdoors and you can even sear a steak over a campfire on the lid. It holds the heat and distributes it evenly. It makes me feel like I'm a little closer to cooking in the way my ancestors did, and that to me is a good thing! The only downside to these is they are very heavy.

Blowtorch

The cheaper ones available are not usually up to the job. You need something with a bit of oomph. A good quality one is super useful for skinning tomatoes, crisping up fish skin, adding a sugar crust to a crème brûlée along with a myriad of other tasks.

Meat Thermometer

If you're prone to undercooking or overcooking meat these are a great help. Ovens can vary considerably, and a thermometer can be a godsend in getting your steaks and, particularly, joints perfectly cooked. Always probe the centre of the meat to obtain an accurate temperature.

Stick Blender

Much more convenient, cheaper and quicker to use than a food processor. It's much easier to clean too, just give it a quick rinse under the tap! We still have one of Mum's that is probably half a century old and it still seems to work perfectly. We use it for all manner of sauces, salsas and soups.

Humble Beginnings

There's no doubt that Mum and Dad were both brought up to appreciate the food on their plates. They were both war babies from farming backgrounds, which carried with it considerable advantages. There was a thriving black market, and food produced on the farm would be exchanged for goods and services with other farmers and businesses. They look back on that fondly, but the situation certainly made them value food that had been difficult to get hold of when it once more became available.

However, there were many things on the post-war menu that didn't tickle their tastebuds. They both remember rationing, and particularly the joy of getting their sweet ration. Our generation has never had to cope with food being unavailable and having to adapt recipes because of that. Times were tough so nothing was wasted and being creative in the kitchen could certainly help.

As I write this, the cost of food is rising sharply and it may be that, once again, we really see the value in the food we eat. Hopefully we will begin to cherish that high-quality, home-produced food in the same way our parents and grandparents did.

Dad's Early Food Memories

There was a time when providing food for the family was a lot more challenging than it is today. Both my mother and father had times when what ended up on the family table, was far from being to their liking.

Dad looks back on his days growing up with fondness, but not always delight at what his mother, Rene, would lay before him on the dinner table. If times were tough, she would take out the old .410 shotgun and shoot a couple of the pigeons that were living in the barn. She would gut and dress them in double quick time and serve them up to the family. They were tough people living in uncompromising times. The dish he remembers with least pleasure was rook pie. Such was the harsh, acidic nature of the bird's flesh, it could turn a silver spoon completely black.

Game was often on the menu, and Dad's favourite dish was jugged hare, he has always loved his food and his eyes shine when he talks about how good it was. He's always been very much a conservationist though, and wouldn't choose to eat a hare now as their numbers have declined so much.

Mum's Early Food Memories

I asked Mum about the sort of food that she ate as a child when growing up on the family farm in Halifax. Theirs was a poultry farm that produced eggs. Times were tough in the post war period, and nothing could be wasted. Mum sat at the kitchen table and related her story with her usual dry wit; delivered with a sideways smile and a chuckle.

'My dad, your grandad, would kill one of the cockerels from the previous year's breeding stock. They were big and tough, you couldn't just roast them in the oven like you can a chicken today. He would put them in a large roasting tin surrounded by water and vegetables. He'd cook them on the kitchen range; the temperature was controlled by how large the fire in the hearth was. They'd take all day to cook, and they smelled bloody awful!' She laughed. 'People must have liked them though, because they would invite them, and they would come from miles around, but I just couldn't stand the smell of it. My mum would sit by the fire and pluck a boiling fowl (a senior hen, well past it's best for egg production), she'd throw the feathers on the fire and that would smell bloody awful, too!'

Even the clear, infertile eggs that had been warmed in the incubator for weeks were used to make sponge cakes. I really can't imagine that it didn't affect the taste!

SPRING

I sit in my garden at Mill Farm and watch the lambs dancing happily in the lush grass around the 17th century miller's cottage that I call home. It's an idyllic scene, and I reflect on the first time I saw this wonderful place when I was a teenager, long before the idea of buying Mill Farm ever entered our heads.

Back in those days, if you had a season ticket for the trout fishing pond up the road, you could also fish the little stream that led all the way down to the lakes at Cannon Hall. My childhood friend, Nick Kent, had one of those tickets and could take along a guest with him fishing. I went with him, and we followed the stream through some pretty rough, prickly undergrowth, doing our best to avoid the brambles and nettles, and not always succeeding.

We came to a little plunge pool under a small waterfall that looked deep... really deep. There was a pleasant coolness in the shadows of the trees that surrounded that secret place as we crouched on a small gravel bar that sat beside the water. We were fishing with worms back in those days. It was a hot day, and the fish hadn't been biting up on the dam, but down here in the shade, the trout were opportunists who would grab anything that plopped noisily into the water. They were very wary though, and even a heavy footstep would see them melt away into the depths, then your chance would have gone.

To our young eyes, that pool looked so deep it could hold a trout of any size. We imagined some old hook-jawed leviathan, sitting in the depths, chomping its way through bullheads, stone loach and crayfish. We dropped our float in the little whirlpool where the waterfall hit the pool's surface and watched as the worm sank into the dark and the float settled.

Almost immediately the float bobbed once, then twice, then slid away, as somewhere beneath the inky surface something had grabbed the wriggling redworm and headed for the roots of a nearby tree. I lifted the rod, and there followed a short but dramatic battle, as the trout twice leapt free of the water landing back with a huge splash. It battled hard for a minute or two, until I managed to coax it over the rim of the waiting net where there it lay – in the folds of the mesh – the most beautiful trout I'd ever seen. It was over a pound, a big fish from that tiny stream, the red spots on its flanks seemed almost to glow, and its belly still gleams yellow, like soft Jersey butter in my memory. In a way I wish I had a photograph of it, but the picture I have in my mind is enough, no image could do justice to that.

After I'd caught the trout, I climbed up the other bank, lay there in the long grass and rested my eyes on Mill Farm for the first time. It took my breath away. I can still remember thinking, as I looked across that hay meadow, what a wonderful little place it was. A cluster of ancient stone buildings, a picture postcard cottage next to what I would later learn was a 16th century mill and an old wooden Dutch barn. It was settled there comfortably in the landscape, with a little wood wrapped around it. Like an old man in his favourite armchair. Like it belonged there. It was beautiful.

Many years later, in 2004, my dad and I were viewing Mill Farm. It had come up for sale after the sad death of the owner Mr Town, who was an old friend of Dad's. We were walking along the banks of that same stream and came upon the pool with its little waterfall, looking much as it had all those years before. As we stood there watching the insects dancing above the water, a small trout leapt clear of the surface, coming down with a splash. My eyes shone bright as I watched it.

'Can we buy it Dad?' I asked, as if I was a small child again.

The lines around his kind eyes gently creased as he smiled at me.

'We'll have a go lad; we'll have a go.'

Now, as I survey that same springtime scene, the barn has gone, but the old mill and buildings are still there, since converted into cottages. The wild garlic is a lush bed of green covering the ground, its white flowers coat the banks of Ronscliffe Dyke. Mayflies dance above the crystal-clear water of the stream and fat little trout still leap from the water. I wonder if they are descended from that long ago fish that so captured my imagination?

Springtime is my favourite time of year, so full of promise and potential. Time to dust off my fishing rods, stock up on charcoal and take a wire brush to that BBQ ready for some al fresco cooking.

Pickled Wild Garlic Buds

Springtime is such a wonderful time of year, even if it does send my allergies into overdrive! New life is bursting out everywhere you look, and one of my favourite sights is a bed of wild garlic, broad, deep green leaves flecked with white flowers. It is unmistakable due to the heady scent of onion that surrounds it.

Down at Mill Farm it coats the banks of the stream in its fragrant bounty. As soon as it appears, sometimes as early as late February, I'll be picking the fresh young leaves to use in pesto and on garlic bread. For me though, the best bit is the flower buds before they've opened. Collected and pickled they make a delicious alternative to capers when sprinkled on a salad or pasta, and while they can take a little while to pick, this is a delightfully simple way to preserve them.

Wild Garlic ...
in cider vin...
16/4/202...

Ingredients

1 cup wild garlic buds*
300ml apple cider vinegar
1 tsp sugar
1 tsp salt
1 tsp pink peppercorns

Method

First sterilise a Kilner jar or jam jar. You can do this using hot water or heating in an oven at 100°C for 30 mins.

Trim the stem close to the bud and wash in cold water, then carefully pat dry with a clean tea towel. Combine the vinegar, sugar and salt in a saucepan, heat and stir until the salt and sugar have dissolved. Add the pink peppercorns and allow to cool.

Place the buds into the jar and cover with the cooled liquid. Put on the lid and leave to pickle for at least a month before using.

Note

These seem to improve with age, making this one foraging exercise that really is well worth the effort!

*   Always make sure your wild garlic has an onion smell to it, as there are some poisonous lookalikes. There's lots of information available online so you can quickly confirm your identification.

Wild Garlic Pesto

Wild garlic is one of the great joys of springtime at Mill Farm. From the time the first vivid green shoots appear, followed by white flowers and then the ransoms (these are wild garlic seed heads that can be preserved, firstly in salt and then vinegar). Both wild garlic and the pesto I make with it freeze well, so you can preserve its delicious flavour and use it well into the autumn.

I prefer this pesto to the classic basil version and there's a huge amount of satisfaction to be gained from collecting your own food and making it into something delicious. A little goes a long way, as it's packed full of flavour. If you're keeping it in the fridge for a few days, an extra drizzle of olive oil on top helps to preserve it.

Ingredients

140g of wild garlic leaves, roughly chopped
1 clove of garlic, roughly chopped
100g toasted pine nuts
100g Parmesan cheese, finely grated
A squeeze of lemon juice
125ml olive oil
Pinch of sea salt

Method

Place all the ingredients into a bowl and using a stick blender, chop until it is the desired consistency. You can choose to chop it finely, or aim for a more rustic version. If you can't get wild garlic just use basil leaves instead.

Bruschetta with Wild Garlic Pesto

Serves 4

Ingredients

4 large ripe vine tomatoes (using different coloured
  heritage tomatoes can create a more colourful dish)
1 small red onion, finely chopped
1 tbsp of balsamic vinegar
3 tbsp of extra virgin olive oil
1 clove of garlic, cut in two
Sea salt and black pepper
French stick cut into 8 pieces
Roughly torn basil leaves

Recipe continues overleaf

Method

Pre-heat the oven to 200°C.

Remove the tomato skins by scorching them with a kitchen blowtorch* until they blister, before slipping off the skin. De-seed the tomatoes and dice the flesh. Add the onion, the balsamic vinegar, 2 tbsp of the olive oil, and plenty of salt and black pepper to taste. Mix thoroughly and allow to sit in the bowl outside the fridge for at least an hour before serving. Tomatoes are *so* much tastier served at room temperature.

Sprinkle the bread with the remaining olive oil and rub the surface with the cut part of the garlic clove. Put the bread in the oven and toast for five minutes, or until the bread begins to brown.

Give the tomato mixture a stir and spoon generously onto the bread. Dot on a little of the wild garlic pesto (see recipe p8) but use sparingly as it's powerful stuff! Garnish with the basil.

Note

* If you don't have a blowtorch, scoring then immersing the tomatoes in a bowl of boiling water for 1 minute will allow you to remove the skins.

Roasted Crushed New Potatoes with Garlic Butter and Stilton

Serves 4

These mini baked potatoes are a delicious, buttery, cheesy side dish when you'd like something a little more interesting than plain boiled spuds. They're very filling so you don't need loads. They're great as part of a tapas style feast or as a vegetarian appetiser.

Ingredients

125g butter, softened
3 sprigs rosemary
3 sprigs thyme
Sea salt and pepper
3 cloves garlic, crushed
Few sage leaves, chopped
750g waxy new potatoes (such as Charlotte)
150g Stilton

Recipe continues overleaf

Method

Preheat the oven to 190°C. Mash the butter in a bowl. Strip the herb leaves from their stems, chop the rosemary finely and sprinkle with the smaller thyme leaves into the butter and season with sea salt and black pepper.

Roughly mix in the rest of the ingredients except the cheese and potatoes.

Boil the potatoes for 18 minutes. Drain and allow to sit for a minute or two.

Place the potatoes on a baking tray and partially flatten them so they crack open. Spoon the mixture over each potato and roast in the oven for 20 minutes.

Remove from the oven, crumble the stilton over and bake for a further five minutes.

Note

I'll sometimes add fried streaky bacon pieces or spring onions to this recipe.

Mum and Dad On Romance

Farmers aren't usually thought of as the most romantic of people. Mum and Dad care deeply about one another, but they certainly don't wear their hearts on their sleeves.

I was chatting with Mum one Valentine's Day. Dad was away, a few weeks from his 80th birthday, buying a bull, in Scotland. 'Getting his priorities right, as always,' Mum said. '61 years ago today, he did get me a bit of something,' she said. 'I've not had anything since. He'd only known me a few weeks, so I must have made some sort of an impression.'

There can't have been many romances before Mum, as they were pretty young when they met, but Mum knew about a few. Apparently, a lass from Bradford way had made something of an impression on young Roger. Mum later recounted the tale: 'He only ever saw her twice. He'd gone down to London with the young farmers, judging a dairy show, when a young lady caught his eye.'

'So, you remember her then, Dad?' I asked him, and he looked at me with a smile, and a twinkle in his eye.

'Remember her?' Mum said, tight lipped. 'He can't even remember what he had for his tea, but he can remember Jenny Booth!'

Mum has the driest and most cynical view of life, very funny when she wants to be. Our family's version of Maggie Smith's Dowager Countess, in *Downton Abbey*. Tough, practical, uncompromising, but full of wit and wisdom, she's always a strong shoulder for the family to lean on, but doesn't suffer fools gladly.

I asked Mum about love, and she said: 'I never tell your dad I love him, and he doesn't me. I think he'd fall over from shock if I ever did! It doesn't mean we don't love each other though. We just don't need to talk about it.'

When Dad drifts off into his own little world, as often happens, and Mum notices he isn't listening to her, Mum will roll her eyes and use one of Dad's mum Rene's many old sayings: 'save your breath to cool your porridge'.

How many old sayings involve food? Quite a lot in our family, that's for sure.

Farmer Richard's Great Balls of Fire Penne Pasta

Serves 4

I cooked up a dish of this in my large paella pan for my 30th birthday party, which took place in what, back in those days, was still known as Home Farm Tearooms. This was the building that would eventually become The White Bull restaurant. In truth, I probably overdid it a bit, chucking in a handful of those hot little Italian birdseye chillies, and afterwards, the DJ, having sampled it during the food break, declared to the watching crowd that 'There was a touch of the Jerry Lee Lewis about that pasta! Goodness gracious great balls of fire!'

I can still remember feeling more than a little alarmed as I spotted six-year-old family friend, William Mott, bright red in the face, with sweat running down his forehead as he battled through my volcanic pasta dish. I've mellowed this version down a little, but feel free to chuck in a few of those little Italian birdseye chillies instead of the Guindillas if you'd like to experience that authentic burn. Leave the chillies out altogether if you prefer a milder but still delicious pasta dish.

Ingredients

300g good quality dried penne pasta
2 tbsp olive oil
150g smoked streaky bacon or pancetta, diced
1 medium red onion, sliced
1 medium red pepper, sliced
100g spicy Italian pepperoni, sliced
8 cloves of garlic, roughly chopped
2 tbsp tomato puree
2 x 400g tins of good quality chopped tomatoes
8 Guindilla chilli peppers (usually found in a jar, pickled) - sliced
Sea salt and black pepper
35g of capers
80g grated parmesan
Handful of torn fresh basil leaves

Method

Put a large pan of salted water on to boil (for the pasta).

Add the olive oil to a large, deep sauté pan, and fry the diced bacon for 4-5 minutes at a medium heat until it starts to crisp round the edges. Add the red onion to the pan and stir for 1-2 minutes until it's starting to soften. Add the red pepper and fry for 2 more minutes. Add the pepperoni and cook for 1 minute then add the garlic and cook for a further minute, continuing to stir.

Add the tomato puree and tinned tomatoes and stir in. I rinse the tomato tins with around 100ml of water, pouring it from one tin to the other and stir the water into the pan – waste not want not, as Gran would have said! Add a good grinding of black pepper and sea salt to taste and reduce the heat to medium low. Simmer gently for around 15-20 minutes, continuing to stir as necessary. Taste the sauce and if the tomatoes aren't as sweet as you'd like, add a teaspoon of sugar. I use good quality tomatoes, such as Mutti or Cirio, so I find this isn't necessary. Add the Guindilla chilli peppers and continue to stir regularly.

Cook the pasta in the boiling water, till al dente (usually 9-10 minutes).

Continue to taste and stir the pasta sauce as it reduces while the pasta cooks. If the sauce becomes too thick, add a couple of tablespoons of the pasta water.

Taste the sauce and season if required. Drain the pasta (saving a little of the water) and add to the sauce, Stirring through until coated, adding a little more of the saved pasta water if required. Serve in bowls, sprinkled with the capers, grated parmesan and basil leaves.

Chicken, Smoked Bacon, Mushroom and Asparagus Risotto

Serves 4

I remember watching my sister-in-law, Marie, making a simple, but delicious, porcini risotto when I was staying with her and her husband Lorenzo in the Italian lakes over 20 years ago. It's amazing how much flavour rice can absorb from good quality wine and stock.

When I returned home, I cooked it myself, adding some of my own favourite ingredients, until I arrived at this version. There's something very special about risotto. There's such a connection with the food, stirring until you reach the perfect consistency. Making a good risotto is almost an act of love, and the more times you cook it, the better you get. I've cooked this recipe many times.

Risottos seem to scare some people, but they are, in fact, very simple. They do require your undivided attention – but isn't that a good thing? Learning to cook a risotto is about feeling your way. If you're not as fond of garlic as I am, then cut back on it. Because I chop roughly, rather than crush the cloves, the garlic flavour isn't as overpowering as you might imagine.

Ingredients

15g dried porcini mushrooms
2 tbsp olive oil
175g smoked streaky bacon or pancetta, diced
1 medium onion, finely chopped
2 chicken breasts, chopped into 2 – 2.5cm pieces
9 cloves of garlic, roughly chopped
300g Arborio rice
Large glass of dry white wine
Black pepper and sea salt
750ml chicken stock
300g white mushrooms – wiped with a
    damp cloth and thickly sliced
1 bunch asparagus, break off base and slice into 1in pieces
120g Parmesan, finely grated
40g salted butter

Method

Add the dried porcini mushrooms to 200ml of boiling water and leave to stand for 15 minutes.

Add the olive oil to a large, deep, non-stick sauté pan and place on a medium heat. It's very important that you stir the dish regularly throughout the cooking process. Fry the smoked bacon for 5 minutes or until starting to brown. Turn the heat down slightly, add the chopped onion and fry for a further 4 minutes, or until the onion begins to soften and become translucent. Add the chicken and fry for 3 or 4 minutes until coloured all over. Add the garlic and fry for a further minute.

Add the rice, stirring well to coat thoroughly with the oil and turn the heat down to a low/medium.

Add the wine and stir for about 5 minutes, or until reduced by half and syrupy in texture.

Add the rehydrated porcini mushrooms and their liquid to the pan. Keeping the heat on medium low, add a few grinds of black pepper and a pinch of salt.

Once the mushroom liquid has almost evaporated, start adding chicken stock, about 125ml at a time. Keep stirring the rice regularly, adding more stock when the previous has nearly all been absorbed.

After about 15 mins of adding stock, add white mushrooms to the pan and continue to stir. The rice should be starting to swell but still chalky to taste. The consistency should begin to look creamy.

Continue to add stock as and when needed. I tend to taste the rice regularly towards the end of the cooking time, as rice can absorb liquid at different rates. After 5 more minutes (roughly 20 minutes in total from the stock first going in), add the asparagus and continue to stir for another 5 minutes.

If you stir using a flat wooden spatula, you'll get a feel for when the risotto is nearly ready. The degree of resistance increases on the bottom of the pan, and the rice starts to stick a little. At that point, taste one final time and adjust the seasoning. There should be a creaminess about the rice and still a little resistance to the bite, but without chalkiness.

Stir in two thirds of the grated parmesan, adjust the seasoning again if necessary and finally stir in the butter for a rich, glossy finish.

Serve immediately topped with the remaining parmesan.

Note

The quantity of chicken stock used during cooking will vary slightly each time you cook it, so have a little more handy, just in case.

Thai Green Fish Curry

Serves 2

The list of ingredients for this one can look daunting and you'll probably have to venture beyond the confines of the local supermarket to find them all, but I always feel like it's worth it. I've found a little Thai supermarket in Sheffield that fits the bill and provides me with everything that's needed. Every time I visit, I come home with lots of things that I don't need, or don't know how to cook. That's one of the joys of cooking, for me, though – discovering the new and adapting it to what I already know. While this may seem like a lot of trouble, the result is *so* much more impressive than buying a paste in a jar. The shop-bought pastes tend to have much less of a flavour profile, sadly, they tend to be hot but one dimensional, lacking the wonderful aromatics of a good homemade paste.

I select the vegetables for colour and texture and often vary them; options include beansprouts, carrots, shiitake mushrooms and mangetout.

Monkfish is great in this curry as it won't break up as it cooks. You can use other fish such as cod or haddock but don't stir the curry or they'll flake up during the cooking. More sustainable and affordable fish, like pollock or coley are delicious in this, and I'll often cook them in a curry after catching them from our boat at Whitby.

Ingredients

Green Curry Paste:

4 tsp Thai fish sauce
20 birds eye chillies
4 Thai shallots, chopped
4 lemongrass stems, chopped
4 tbsp galangal, chopped
I tbsp shrimp paste
Zest and juice of two limes
2 tsp ground coriander
2 tsp ground cumin
½ tsp white pepper
1 tsp turmeric
8 tbsp chopped coriander (leaves, stems and root if available)
6 lime leaves, shredded
8 garlic cloves, chopped
6 tbsp vegetable oil

Method

Combine the ingredients and make into a smooth paste using a hand blender or food processor. This paste provides the basis of any Thai green curry. Although it does take a fair amount of effort it freezes well, so I make a large batch and freeze it in individual portions. If you enjoy your curries hot, add an extra spoonful or two of the paste to the dish.

Recipe continues overleaf

Ingredients

Fish Curry:

600g of monkfish fillets cut into 1.5in pieces
Juice and zest of a lime
Sea salt
2-3 tbsp groundnut oil
2-3 tbsp Thai green curry paste
6 Thai baby aubergines, chopped into 1in pieces
1 red bell pepper, chopped into 1in squares
125g baby sweetcorn, sliced down the middle
    if large, or left whole if small
1 tin coconut milk
6 kaffir lime leaves
1 tsp fish sauce
2 tsp light soy sauce
100g sugar snap peas
1 red chilli, chopped
2 spring onions, chopped
Handful of chopped coriander

Method

Marinate the monkfish in the lime zest, half of the lime juice and a sprinkling of sea salt for about 20 minutes.

Heat the groundnut oil on a medium heat in a large flat-based frying or sauté pan, add the Thai Green curry paste and fry for two minutes.

Add the aubergines and fry for 5 minutes. Aubergines can soak up a lot of oil, so add a little more if needed.

Add the red pepper and baby corn. After 2 more minutes add the coconut milk and stir. After 1 more minute add the kaffir lime leaves, fish sauce and soy sauce and stir.

Add the monkfish pieces to the pan, ensuring they're submerged into the coconut milk to poach. Lower the heat to a gentle simmer.

After 5 minutes add the sugar snap peas, pushing down into the coconut milk.

Depending on thickness, the fish should cook in approximately 12-18 minutes.

Finish with the rest of the lime juice, it retains its freshness when added at the end. Sprinkle with the spring onion, red chilli and coriander.

You can serve with Thai sticky rice, jasmine rice or the Super Smashing Great Egg Fried Rice (see page 35).

My Dad

Roger Nicholson is a special man. Everyone thinks their dad is special, I know that, but Roger is a unique human being. I've never known anyone as quietly tenacious and single-minded as he is in his approach to life, and to looking after the animals in his care.

I was reminded of this recently, when he showed me a very poorly Swiss Valais lamb that he'd taken on as a project. He was trying to nurse it back to health and looking after it in a loose box, close to his house. These boxes are one of the last remaining parts of the original farmyard set on one side of the square that now forms Mum and Dad's garden. Back in the 1970s this area was covered by a large, iron framed roof. During one particularly bad winter, the snow kept coming and that roof buckled under its weight. I remember seeing it, through a child's eyes, half collapsed, the metal beams crumpled up. I never considered the financial implications for the place, and of course Mum and Dad kept that from us. It was another nail in the coffin for them. There was no way he could afford to rebuild it. It was just another disaster in a long string of difficulties that life had thrown at him.

Stoic doesn't even begin to cover what this man is about. My child's eyes have seen the farmyard littered with the bodies of cattle, legs pointing skyward, stomachs swollen by gas, victims of poisonous rhododendrons they'd eaten during a spell of wintery weather when the grass they'd normally rely on was covered in snow.

We watched our parents struggle to make ends meet throughout our childhood and teenage years, but they always delivered us thoughtful presents at Christmas and on birthdays. All the while they were watching the debts slowly mount up. We fully appreciate it now, but never really understood the pressure they must have been under at the time.

Farmer David's Sticky Chicken with Super, Smashing, Great Fried Rice

Serves 4

For Farmer David's favourite recipe we take a trip to the Orient. Juicy chicken thighs, coated in panko breadcrumbs, shallow fried and covered in a hot sticky sauce. Gochujang is one of my favourite ingredients, a fermented chilli paste from Korea that you can buy in most supermarkets. It can all get a little messy, and my kitchen does look like a bomb's hit it when I've cooked it, but I think it's worth it! It's lip-smackingly tasty, and so much nicer than a takeaway.

Ingredients

Marinade:

2 cloves garlic, crushed
2 tsp soy sauce
2 tsp rice wine
1 tsp chilli powder
¼ tsp salt
¼ tsp white pepper

Chicken dish:

6 boneless skinless chicken thighs, cut into 1in pieces
100g cornflour
1 large egg, beaten
120g panko breadcrumbs
Sunflower oil for shallow frying
3 tbsp groundnut oil
1 red pepper
150g of baby sweetcorn
100g mangetout
4 cloves garlic crushed
2 tbsp ginger, grated
85g runny honey
50g brown sugar
2 tbsp soy sauce
1 tbsp rice wine vinegar
1 tbsp sesame oil
200g fresh pineapple, ½in diced
2 tbsp gochujang
4 spring onions, chopped
1 tbsp sesame seeds, toasted
1 red chilli sliced

Recipe continues overleaf

Method

Mix the contents of the marinade together, coat the chicken pieces and marinate for 1 hour.

Place the cornflour, egg and panko breadcrumbs in separate bowls. Dust the chicken pieces in the cornflour, then dip in the egg and roll in the panko breadcrumbs.

Fill a large deep sauté pan with sunflower oil to about ½ an inch in depth. When the oil is up to a medium-high temperature, cook the chicken in batches for about 2-3 minutes each side. Cut one open to check it's cooked all the way through. Place on kitchen paper and keep warm.

Heat the groundnut oil to a high temperature in a large frying pan or wok. Cook the red pepper and baby sweetcorn for 3 minutes, add the mangetout and cook for a further minute, then add the garlic, ginger, honey, brown sugar, soy sauce, rice vinegar, sesame oil, pineapple and gochujang. Stir and heat through until it bubbles.

Add the chicken back into the pan and coat with the sauce until warmed through.

Sprinkle with the spring onions, sesame seeds and sliced chillies. Serve with Super Smashing Great Fried Rice (see page opposite).

Super, Smashing Great Egg Fried Rice

Serves 4

We were all big fans of *Bullseye* as kids! We loved a bit of Bully, and if Jim Bowen ever cooked fried rice (which he probably never did) this would be the rice he'd make, (possibly)! Vary the chopped vegetables depending on what you fancy.

Ingredients

40g butter
3 eggs
1 small onion, finely diced
1 carrot, finely diced
3 cloves garlic, crushed
800g cooked long grain rice, chilled
100g small, cooked prawns
2 tbsp oyster sauce
3 tbsp soy sauce
1 tbsp sesame oil
4 spring onions, finely chopped

Recipe continues overleaf

Method

Heat half the butter in a sauté pan using a medium heat. Scramble the eggs until cooked through. Remove from the pan and keep warm.

Add the rest of the butter and fry the onion and carrot for 4 minutes, until both are becoming soft and the onion is translucent.

Add the garlic and stir for 30 seconds. Add the rice and a little more butter if needed. Using chilled rice helps to keep the rice separate and avoid clumping. Using butter helps to brown the rice. Cook for 2 minutes then add the prawns and heat them through for a further two minutes.

Stir in the oyster sauce, soy sauce and finally the sesame oil.

Serve immediately, sprinkled with the spring onions.

Farmer David's top tip: Pair this with a Lucky Buddha lager!

Monkfish Wrapped in Parma Ham with Roasted Red Peppers and Feta Baked Potatoes

Serves 4

This is one of those recipes where you can get most of the preparation done before, so all you need to do is pop it in the oven, and chat with your dinner guests over a glass of wine while you wait.

Ingredients

2kg monkfish tail
100g Parma ham
125g butter, softened
3 sundried tomatoes, finely chopped
Zest and juice of one lemon
2 tbsp chives, finely chopped
2 tbsp parsley, finely chopped
Ground black pepper

Recipe continues overleaf

Method

Preheat the oven to 200°C. Remove the central bone from the monkfish taking a fillet on each side, they are a cartilaginous fish, a member of the shark family, so they only have one central bone. Trim off the grey membrane that covers the fish.

Lay the Parma ham out on a sheet of cling film and place the monkfish fillets onto it.

Mash together the softened butter, sundried tomatoes, lemon zest and juice, chives, parsley and black pepper to taste. Place the butter in a long thin sausage on the middle of the top of the fish. Roll the Parma ham around the fish to form a sausage using the cling film and securing with wooden skewers. Place in the fridge to allow the butter to firm up.

Remove the cling film to cook and bake in the oven for 20-25 minutes, depending on the thickness of the fish. Remove the skewers and slice the fish into portion sized pieces and pour over the juices from the cooking tray.

Baked Crushed Potatoes with Olives and Feta

Ingredients

500g waxy new potatoes
75g butter
Sea salt and black pepper
Small red onion sliced
Handful of black olives
4 cloves of garlic, roughly chopped
3 rosemary sprigs, leaves stripped
200g feta cheese, crumbled

Method

Preheat the oven to 200C.

Boil the potatoes for 20 minutes. Drain, then half crush them with a masher but don't break them up.

Add the butter, salt and pepper and allow them to sit in the pan until the butter melts, then gently fold it all together so the butter soaks into the potatoes.

Layer the potatoes in a baking dish and sprinkle over the red onions, olives, garlic and rosemary. Top with the feta cheese and bake on the top shelf of the oven for 25-30 minutes.

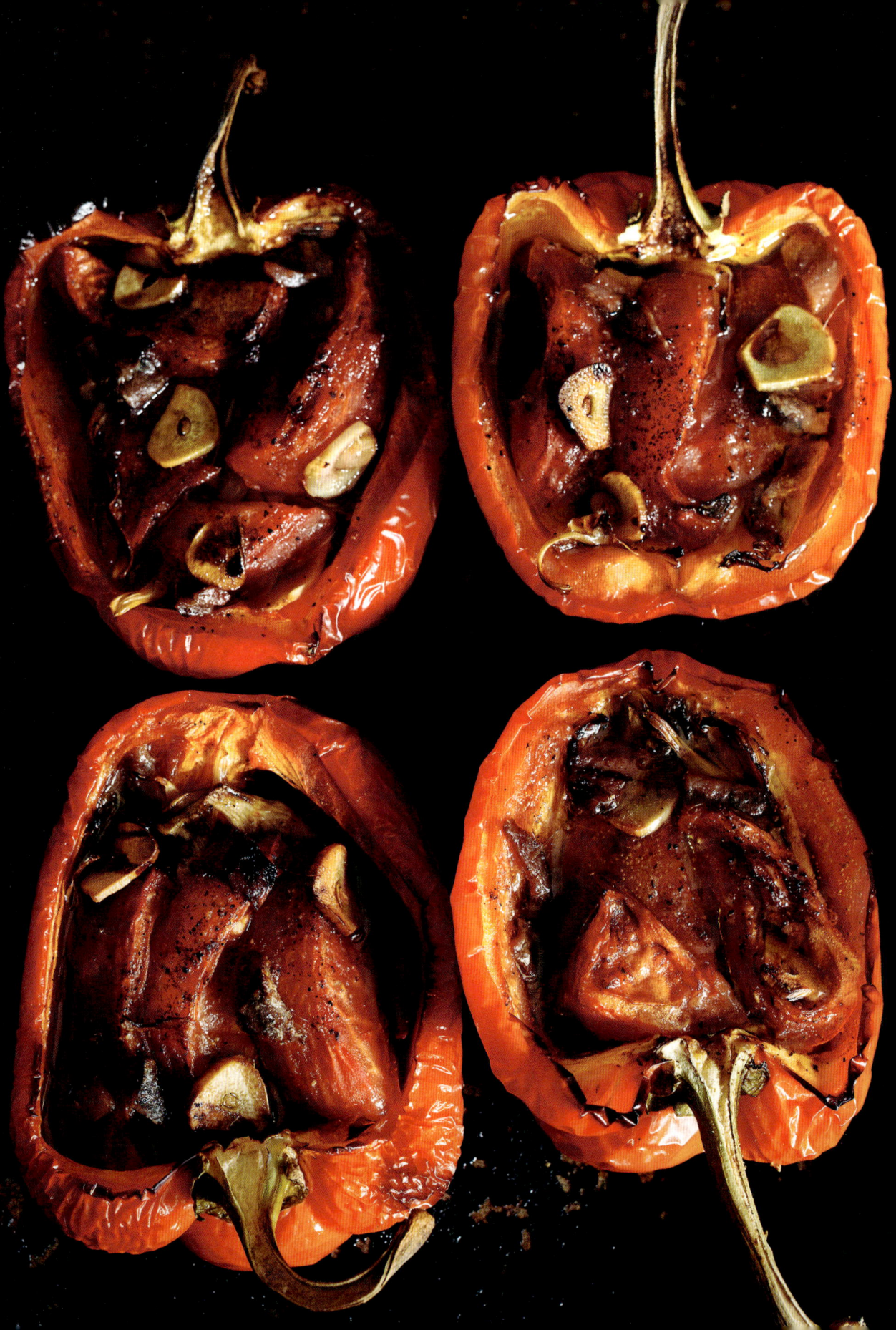

Roasted Red Peppers

Ingredients

Two red peppers
2 large vine tomatoes
½ small red onion, thinly sliced
½ bulb fennel, thinly sliced, fronds reserved
2 cloves garlic, sliced
4 anchovy fillets, chopped
4 tbsp olive oil
2 tbsp balsamic vinegar
Black pepper
Basil leaves to serve

Method

Preheat the oven to 180°C. Cut the pepper in half through the stalk. It looks so much more attractive on the plate when the stalk is sliced in half. Remove the seeds and pith, being careful not to cut through the skin of the pepper.

Skin the tomatoes by blistering using a blowtorch or score the bottom of the tomato with a cross and cover briefly with boiling water before removing the skin. Layer the slices of onion and fennel in the bottom of the pepper, adding small pieces of garlic and anchovy. Add quarters of tomato and more garlic and anchovy pieces. Pour a tablespoon of oil and half a tablespoon of balsamic vinegar over each pepper half and finish with a grind of black pepper.

Bake in the oven for 1 hour. Garnish with fennel fronds or basil.

The Scones That Saved The Farm

Makes 24 scones

Once we had decided to open a tearoom, we had to choose what we'd sell in it. The foundation of a good tearoom is, of course, an excellent homemade scone! Mum and her friend, Rosemary, spent many long hours trying different recipes, my brothers, my dad and I being highly enthusiastic testers! They wanted a scone that would sit up and look the part, not lose shape on the baking tray. It had to slice evenly and give a nice firm base for the jam and cream. In the end, they came up with their own version, and it has served us well for 40 years. I prefer them just warm from the oven, with a generous coating of butter.

Ingredients

250g margarine
880g self-raising flour
250g caster sugar
150g mixed fruit
2 bantam eggs (or 1 hen's egg)
325ml whole milk
1 egg, beaten (for glazing)

Method

Pre-heat the oven to 240°C.

Rub the margarine into the flour until you get a fine crumb texture.

Recipe continues overleaf

Using a food mixer with a dough hook, quickly mix in the rest of the ingredients (except the beaten egg). It's important not to overwork the mixture at this point to obtain a result with a light texture. Roll out the mixture to about an inch thick and cut out your scones. If you roll out the mixture again, bear in mind that these won't be quite as light as the first batch you've cut.

Glaze the scones with beaten egg.

Divide between two baking trays and put one batch on the top shelf and one on the lower. Bake for 5 minutes at 240°C, then rotate and swap the trays between the top and bottom shelves, lower the temperature to 180°C and cook for a further 10-12 minutes.

Note

This seems like a rather complicated way to do things, but it was the way Mum did it. She tells me that every oven is different, and every time she changed her oven, she's had to adjust the temperatures slightly.

The original recipe called for two bantam eggs: a type of small, domesticated chicken that lays smaller eggs. Most hen breeds have a smaller bantam variety. Not easy to get hold of unless you happen to keep them. The recipe works perfectly well if you substitute with one hen's egg. Mum didn't bother with caster sugar and used granulated, but I prefer to make them with caster sugar.

Home Farm Tearooms opened to the public on Good Friday 1981. So successful was this recipe that large quantities were required every weekend in our little tearoom. Mum would rise in the early hours of the morning on a busy summer Sunday, or Bank Holiday and would 'have to bake 13 dozen of the bloody things!'

SUMMER

A long, hot, Mill Farm summer is something to be treasured. In the cool of the morning, as the mist rises from dew-coated grass, I'll sometimes spot a roe deer from my bedroom window, highlighted by the day's first rays of sunshine as they filter through the tree branches, and bathe the meadow in front of my cottage in dappled light.

After weeks without even a shower, the stream can slow to little more than a trickle, and the trout will hunker down in the deepest pools and wait for rain.

It's a time when parties start in the afternoon and extend into the warm evenings. A time when sizzling summer barbeques, and al fresco eating are the order of the day. A chance to lose yourself in the slower pace of life that Mill Farm offers. Sitting in the garden with friends, cradling a glass of wine and putting the world to rights as bats dart to and fro above us, in search of insects in the twilight. Blankets wrapped around our shoulders, we sit around the firepit and watch the glowing embers slowly die.

Visitors From Across The Pond

Our family has spread far and wide over the years. Dad's mum, our gran, made friends in many far-flung parts, and this has led to visits from all over the world. From the Netherlands, New Zealand, and Sri Lanka. One year, our Auntie Flo's daughter, Norma, came to stay on the farm, along with her daughter Jane. Norma had moved to Chicago when she was young, met her husband Emile, and they'd settled down and raised a family. They were friendly, larger than life characters and they loved being back with family, in the old country.

There were many memorable moments during their visit but the one that sticks in the memory was the time when Norma was in the bathroom that overlooked the farmyard. It was broad daylight; she heard a commotion outside and poked her head out of the window. There was much squawking, screeching and flapping of feathers. As Norma watched, the farmyard's flock of hens and bantams stampeded into view, a multi-coloured Heinz 57 of the avian world, winged mongrels of questionable parentage. Behind them, closely pursuing, snapping and salivating and desperate for his next meal, was a very hungry fox.

You may think that this sort of thing happens on farms all the time, but far from it! Your average fox is a timid opportunist who skulks around in the shadows and usually only ventures out at night, but this fox hadn't read the script! Norma watched him hightailing it up the yard in hot pursuit of our rag tag band.

Later we checked for clumps of feathers and counted all the birds – luckily, they'd managed to stay out of the foxy whiskered gentleman's reach.

Back in the house, as the strange and colourful stampede disappeared up the farmyard, Norma was heard to shout, 'Gee! It's just like the movies!'

What follows is a recipe straight from the States that uses Campbell's condensed soup, the American classic, immortalised in that well-known painting by Andy Warhol.

Carrot Salad

This is an American recipe that is often called copper pennies. Mum's friend Rosemary served it up at a party, and I loved it so much I asked for the recipe. It became a regular at our New Year's Eve parties. It's a sweet and sour salad – a great buffet dish. It goes brilliantly alongside smoked salmon or cold cooked meats. I can't understand why it's never become popular in the UK. Perhaps it's that quirky list of ingredients. I love the taste of this though!

Ingredients

900g carrots, sliced
1 green pepper, sliced
1 onion, sliced
1 tin Campbell's condensed tomato soup
150g sugar
180ml red wine vinegar
120ml olive oil
1 tbsp Dijon mustard
1 tsp Worcestershire sauce
1 tsp fresh dill or ½ tsp dried dill

Method

Boil the carrots for 5 minutes, drain, and place in a bowl to cool before adding the pepper and onion. Mix the rest of the ingredients in a saucepan and heat gently on the hob – making sure it doesn't boil – while whisking until all the ingredients incorporate.

Allow to cool, then pour over the carrots, onion and pepper. Cover and refrigerate for at least 24 hours to allow the flavours to develop.

Auntie Shirley and Uncle Laurie (Mounty)

Mounty was my uncle, Auntie Shirley's husband, Laurie. He was as Barnsley as they come. If ever there was a bloke that was made for the phrase 'God broke the mould when they made him', it was Laurie. He was that sort of man. A maverick, a complete one off.

Laurie and Shirley's house was very much a home of the 1970s... The smell of the place – it was overwhelmingly redolent of tobacco smoke, and there was constant cooking going on, both on and in the AGA. It had a large picture window looking onto a rotting concrete and iron balcony, that we weren't allowed to walk on due to mining subsidence from Redbrook Colliery next door. Over engineered ashtrays on stands spun at the press of a button, to deposit cigar ash into the receptacle below. The coffee table was huge, made from an epic chunk of gnarled wood, cut straight from the base of a tree.

Laurie's son, our cousin Alwyn, drove around with a pair of steer's horns on his pickup and a horn that played a tune, like the one from *The Dukes of Hazzard*. He had a look of Burt Reynolds and was very popular with the ladies. His twin, Richard, couldn't have been more different. He was quiet, thoughtful and artistic. Alwyn was his father's son, a charismatic but unpredictable character and Richard much more like Shirley, his shy, subdued and unassuming mum.

Guests to the household included world ferret legging champion Reg Mellor, and well-known local character Ken Hadfield, who once upset Mum by suggesting her scones would make good ammunition for the Falklands War.

They would sit around a small table in the kitchen, surrounded by a fug of cigar smoke, drinking whisky and playing cards. Laurie, it was said, was ruled by the moon. A more unpredictable character you couldn't imagine, he had tousled hair, wild eyes and curved yellow teeth that parted into a snaggle-toothed grin. The shelf above the bathroom sink had a tube of smoker's toothpaste on it, though it seemed to have little effect on Laurie's fierce nicotine-stained tusks. Brylcreem was always on there, too, and I'd sometimes pinch a dollop of it and comb it through my hair, although I never achieved the teddy boy quiff I was hoping for.

Upstairs was a huge room that took up the whole top floor. It had a full-size snooker table and bookshelves crammed with Alwyn and Richard's cowboy annuals from the decade before. As kids, my brothers and I loved that room. The whole house had a unique smell about it, and that room in particular smelled, in a quiet way, of adventure. Their old farmhouse stood on what is now a housing estate, but I can't drive past without remembering what it once was: a house full of character and life. I remember soda syphons on a 1970s sideboard bedecked with whisky bottles; shag pile rugs decorating the floor. Fierce looking Alsatians and furious, bug-eyed chihuahuas guarded the doorway at the top of rough concrete steps. I always felt like an innocent abroad in that house.

Alan, one of our first butchers in the farm shop worked for Laurie and Shirley back in those days. He joined them at 16 years of age, working first at their butchery concession in Simco supermarket on Shambles Street in town, and then later in their farm shop, which was underneath their farmhouse at Redbrook.

Barnsley football club's captain, Oakwell favourite Eric Winstanley, was hired to open Laurie's butchery concession at Simco, with the promise of a meat hamper to take home. Having done his bit, Eric was gifted a generous box of goodies, including some excellent steaks, sausages and joints. The next week, Eric turned up at the counter again, ordered lots of meat, gave a cheery wave and walked off with it. This was not part of the agreement as far as Laurie was concerned, as he considered he'd been well paid in the first week, and he was more than miffed.

Eric turned up the following week, too, and stood at the corner of the counter, smiling, and giving young Alan a cheeky wink. Laurie clocked him straight away! Every time Alan made to move towards him, Laurie, watching the situation like a hawk, would growl out of the side of his mouth into

Alan's ear 'Don't you dare serve him!' I believe his fatherhood may have been questioned at some point, too.

I'm told Eric stood there for a very long time, smiling and winking, cheekily, before he got the message and walked away, looking a little crestfallen.

Alan told us many a story, most of which I couldn't repeat, but, safe to say, Laurie was a clever, talented, interesting, funny, totally unpredictable man, who could be all seasons in one day. As a child you'd walk through the doorway into the house and see him there, staring at you with striking eyes from below wild hairy brows and wonder what on earth was going on in his head today.

Laurie bought me my first ever pint in a pub. He really shouldn't have, but rules never seemed to apply to Laurie. I think I was about 14. We'd been to a football match, Sheffield United away I think it was, on a match day special coach. We made it back to the Tom Treddlehoyle pub in Pogmoor just before closing time. We streamed off the bus and into the pub. It was busy and noisy as people chatted loudly about the match, the air was heavy with smoke and the heady smell of hops from the popular Barnsley bitter that was brewed locally at Oakwell. Laurie looked me up and down, said 'you'll do,' and ordered me a pint of Guinness. I took a sip and couldn't believe that people actually drank this stuff voluntarily! I grimaced. I didn't like the taste at all, but struggled manfully to finish it so as not to lose face. Relieved, I managed to drain the last foaming dregs. Then much to my dismay, he ordered me another!

He loved Shire horses, and in his later years he restored an old wooden wagon designed to be pulled by them. I remarked at how impressed I was that he'd been able to complete such an amazing restoration. He studied me a moment, with a look of mild amusement on that untidy face of his and told me you could do anything if you set your mind to it. Well, those were not his *exact* words. Ask me about the exact words if you ever have a chat with me over a few pints.

He's still a bit of a Barnsley legend for those of us that knew him. We walked behind his coffin to the church. It was carried by the wagon he rebuilt, pulled by two of his Shire horses. His death came sadly and predictably, through lung cancer, in his early seventies. But his life was never boring.

The Long Hot Summer of '76

I have vivid memories of the summer of 1976. The rhododendrons that grow next to the lawn behind Cannon Hall seemed to be in bloom forever that year. We spent hundreds of hours playing football on that lawn, being periodically chased off by the head gardener Mr Hales. He looked after that grass better than the Wembley groundsman – and this was our Wembley! We were the bane of his life – and he was the bane of ours! He didn't even need to shout at us! The cry would go up! 'Halesy's coming!' He'd emerge from the archway that led down to the courtyard, a stooped, brooding figure, with an air of malevolence. We'd grab our ball and the T-shirts that had been our goalposts, and we'd run to one of our many secret dens in the bushes.

A glass of Mum's homemade lemonade was just the ticket, as we cooled down after an epic football match – or one of those heroic escapes.

Homemade Lemonade

I love fresh lemonade, and for me it's a drink made for parties in the garden on long, hot, summer days. The acidity of lemon juice is counteracted with sugar, leaving us with a delicious tangy drink that's perfect to quench the thirst.

As kids we all read an awful lot of Enid Blyton books, and they often seemed to be punctuated by a high tea on a warm summer's day washed down with lashings and lashings of delicious homemade lemonade, or ginger beer. I loved those stories. In essence, some were pretty close to the childhood we lived, but the children usually managed to have far more exotic, interesting and dangerous adventures. I loved rereading The Adventure Series to my son, Marshall, as he was growing up and was surprised how much I remembered, especially from *The Castle of Adventure*.

As lemons will vary in their sourness, this recipe is just a guide. You should always taste your lemonade and then add more water, sugar or occasionally lemon juice. If you can get hold of them, Sorrento lemons make wonderful lemonade – but they're sweeter, so you may not need quite so much sugar.

Ingredients

200ml sugar
750ml water
250ml lemon juice
Lemon slices thinly cut
Mint sprigs

Method

Add the sugar to 250ml of the water and bring to a simmer, stirring until the sugar has dissolved. Allow to cool.

Add the lemon juice and the rest of the water. Stir well and taste. Add more water if you feel you'd prefer it more dilute, add a little more lemon juice if you feel it's too sweet, or a touch more sugar syrup if you feel it's too sour. Bear in mind that melting ice will dilute the mixture further if you add it to the jug.

Serve up with the lemon slices and sprigs of mint for a sublime summertime treat.

My Time as a Tie Salesman

The village green was where it all happened. It was the Summer Village Fete, and an opportunity to raise money for the church and local charities. Give an eight-year-old Richard a job selling something he believed in and a firm business vision, and I was your man!

Ron Carbutt has been Mum and Dad's friend for as long as I can remember, he's a village stalwart and champion of many a good cause, a really lovely fella. He could see my potential. He called me over; he'd spotted that I was a born tie salesman! He collected several ties off a clothing stall and hung them over my arm. There were thin ones from the 1960s and super fat, extravagantly patterned ones from the 70s, I think one even had a gravy stain or two on it – after all, this is Yorkshire!

I set off with gusto. 'Ties for sale, 2p each!' I bawled loudly at passers-by. So taken were they with my enthusiasm that they started buying them, and in double quick time I had raised 8p. I even managed to sell the one with added gravy. I went back for more stock as my remaining ties looked rather sad. They didn't have any more, but I was pleased I'd done my bit, and Ron seemed delighted with the 8p.

My career as a tie salesman started and ended that day, but I know it's something I can always fall back on if I need to!

Here's a recipe that would go down well at a village fete – a delicious elderflower cordial.

Elderflower Cordial

I tried making elderflower cordial for the first time a couple of years ago and was delighted with the results. I gave one of the bottles to Dad; he loved the taste, and he'd soon finished the lot. It's wonderful to gather ingredients from the wild and make something delicious from scratch. Depending on where you are in the UK you'll find elderflowers in bloom in late May and throughout June. Collect the flower heads on a dry, sunny morning, as their aroma fades in the afternoon. Try to collect your blossoms away from roadsides if you can, so that your flowers are pollution free. Choose the ones with fresh white flowers and nip off the flower heads just below where the flowers branch out, keeping the stem as short as possible.

Check for any insects before using, but don't wash as this will remove the pollen that will help to give your cordial its unique taste. You should be able to get citric acid online or from a brewer's shop.

Ingredients

1kg granulated sugar
1.5 litres water
20-25 fresh heads of elderflower with
    white flowers, stalks trimmed
2 unwaxed lemons, skin zested and sliced into rounds
30g citric acid

Method

Put the sugar and water in a large saucepan or stock pot. Heat very gently and stir regularly until the sugar dissolves. Do not boil.

Once the sugar has completely dissolved, bring the pan to the boil, then take it off the heat.

Check the flower heads and get rid of any flowers that are dirty. Remove as much green stalk as possible, while still keeping the flowerheads intact, too many green stems can make your cordial bitter. Try not to loosen any pollen .

Transfer the flowers, lemons, lemon zest and citric acid into the sugar syrup. Cover the pot and leave to infuse overnight.

Line a colander with muslin cloth and place over another large pan or bowl. Ladle in the syrup and allow it to pass slowly through the cloth.

Using a funnel, transfer the strained liquid into sterilised glass bottles. You could run glass bottles through the dishwasher to sterilise them, or wash well with soapy water, rinse, then leave to dry in a low oven.

You can drink the cordial straight away, or it can be kept in the fridge for up to six weeks. Alternatively, it's a great idea to freeze it in ice cube trays and use as needed. Mix it with sparkling water for elderflower pressé or add to wine, prosecco or champagne to make a delicious summertime party drink.

It's The Future!

It was on a visit to the home of my old college friend, Stewart Owen, in Keighley in the mid-1980s that I discovered the revelation that was garlic bread – to the soundtrack of Wham's *Last Christmas*. What a memorable evening it was! We went to an icy Howarth and watched cars sliding around the pub car parks, bumping into each other.

Our first call of the evening was to a little Italian restaurant in Keighley. Stewart knew the owner well and ordered a garlic bread starter. I was very suspicious when it came out. After all, I was a lad from the country, not used to posh foreign food, with a dad who regarded anything that wasn't meat and two veg with a deep suspicion that verged on paranoia. Stewart persuaded me to take a bite, promising me a moment of revelation.

Tentatively, I put a piece in my mouth. I crunched through the outer layer, a picture of trepidation. I tasted the rich, creamy, garlic butter, smothered on a crispy exterior – and my mouth was suddenly a blissful place! I was in heaven! Peter Kay would describe a similar moment many years later, and I have to say, he was absolutely right, garlic bread was indeed the future, and I was captured, hook, line and sinker! I've loved and embraced all Italian food ever since.

A Mill Farm barbecue isn't complete without this served up as one of many side dishes. My niece Katie declares it 'the best garlic bread in the world!' I'm not sure about that, but it always hits the spot alongside a plate of succulent, barbecued, farm shop meat.

Wild Garlic Bread

Don't panic! You can make this using ordinary garlic and parsley if the wild stuff isn't available – but as it grows here in such profusion it's lovely to use when in season. Wild garlic freezes well too, so you can preserve it for later in the year.

Ingredients

1 bulb garlic
Sea salt
Black pepper
Olive oil
175g Jersey butter, softened
3 cloves garlic, roughly chopped (if you like your garlic
    butter mild, leave this out, remember you're in charge!)
Small bunch wild garlic leaves, or flat leaf parsley, finely chopped
Squeeze of lemon juice
1 farm shop sourdough loaf or a French stick

Method

Slice the bottom off the whole garlic bulb, season with salt and pepper, drizzle with olive oil, wrap in foil and bake at 180°C for 40 mins. When soft and squishy, remove the garlic from the foil, squeeze out the contents and discard the papery part of the bulb.

Soften the butter. Then add the roasted raw garlic, the wild garlic or parsley (and the chopped garlic if using) to the butter. Add a squeeze of lemon juice and mix well. Season to taste. Slice the sourdough and crisp in a toaster, or put it in the oven for 5 minutes. Smear on the garlic butter liberally and bake at 200°C for a further 9 mins.

Katie and The Dragonfly

My niece, Katie, was about eight years old when she first came fishing with me in the lake at Cannon Hall Park. I'd spent many long hours fishing there as a child, and I knew those lakes like the back of my hand.

It was a lovely, early-summer day, and we were catching quite a few small fish underneath the waterfall that runs into the bottom lake. We were having a lovely morning in the warm sunshine, and little rudd and roach were swirling and splashing on the surface amongst the lily pads, when something else caught my eye. A few yards from the bank, caught in the surface film, sending out gentle vibrations across the water with its four broad wings, was a huge dragonfly, becalmed and desperate.

It seemed such a sad way to go, slowly drowning in the oily surface. We decided we'd try to rescue it. I'd come in trainers rather than anything waterproof, and the landing net we had just didn't quite reach. Katie had some wellies on, but they were only little. She shuffled out into the water, a determined look on her face, holding the landing net at arm's length. I had visions of her falling in headfirst and me having to take her home to her mum, Julie, wet through and covered in pond weed and mud.

Katie has always been a tenacious character, and she was intent on reaching the insect which was still quivering on the lake's surface. The water was so close to spilling over the top of her wellies, my heart was in my mouth. I kept asking her if she was ok, all the while half expecting to see water begin to gush into her boots and watch her topple headfirst into the lake. Somehow, amazingly, she reached out a few extra inches, and the massive dragonfly slipped over the top of the net. She pulled it in gleefully, smiling in triumph, and making her way back to the bank. The insect lay there in the folds of the landing net, its iridescent colours dulled.

Was it already too far gone? Could it survive?

We carefully removed it from the net and placed it on top of a nearby plant. It seemed undamaged, and with its four huge wings outstretched, it was larger than my hand. A mini dragon, shining in the sun, a fearsome apex predator of the insect world. For several minutes it just sat there feeling the warmth of the sunshine, its wings occasionally flickering, its shimmering blue colours slowly returning. After about 15 minutes, its wings whirred into life, and it rose into the sky and darted gleefully off. We could have cheered! Neither of us have ever forgotten that dragonfly.

It was underneath this same waterfall that I caught my first brown trout. It was a beautiful fish with a yellow belly and bright red spots. I always return wild brown trout now, but back then they were a trophy and a feast, although they were always a little muddy tasting from the park lakes.

Like any fish, the fresher they are, the better. This next recipe for trout pate has been a regular on the Cannon Hall Farm party menu and is lovely served on crusty bread as part of a summer picnic.

First Catch Your Trout

Fly fishing is one of the great loves of my life and a successful trip means fresh trout is on the menu. I love the idea of fashioning a fly, with feathers and thread, that imitates an insect well enough to convince a trout to eat it.

If you're buying your trout from a fishmonger, a fresh fish will have firm flesh, bright eyes and no fishy smell. The softer the flesh and cloudier the eyes, the longer it's been hanging around – so it's best avoided.

In the warmer months, I'll often bring a trout or two back from a fishing trip. I like to cook the fish in foil, as this prevents it from drying out and creates a softer, smoother pate.

Homemade Mayonnaise

Why not make the mayo for the trout pate yourself? It's simple and a much more natural product than shop bought, which is laced with emulsifiers and thickeners. Use a neutral tasting oil, such as sunflower. Don't use extra virgin olive oil, or it will dominate the taste of your mayonnaise. Also, don't keep your eggs in the fridge – there's no need. Eggs are designed to remain fresh at normal temperatures, while the bird is laying the rest of their eggs. You can use a whisk or hand blender. It goes without saying the mechanical way is easiest, but if you're working on your biceps and forearms, whisking will help!

Ingredients

1 egg yolk
1 tsp Dijon mustard
240ml oil
½ tsp lemon juice
½ tsp white wine vinegar
Salt, to taste

Method

Mix the egg yolk and mustard with your whisk or blender, then add the oil a drop at a time as you continue to whisk. As the mayonnaise begins to come together you can add a little more each time, but it can split if you get too enthusiastic.

When all the oil has been added, and the mayonnaise has thickened, whisk in the lemon juice and white wine vinegar and season to taste.

Use it as part of your trout pate. It will keep for a couple of days.

Trout Pate

Ingredients

1 or 2 whole trout to make about 900g in weight, gutted and
    cleaned. This will give you roughly 450g of cooked trout.
Juice of 1 lemon, plus half a lemon, sliced
30g butter
120ml Cottage Cheese
90ml mayonnaise
120ml double cream
½-1 clove of raw garlic, crushed *optional*
Plenty of freshly grated nutmeg
Salt and black pepper
French stick, sliced
Sprigs of dill to garnish

Recipe continues overleaf

Method

Take the trout and lay it on a sheet of foil. Squeeze over the juice of half a lemon and dot some butter on top and inside the cavity, along with the lemon slices.

Season with salt and pepper then fold the foil into a parcel and bake at 180°C for 30-40 mins, or until the fish lifts away from the bone. Allow to cool, then separate the flesh from the bone and break the fish into pieces.

Mix together the remaining ingredients, then stir in the flaked trout and mix thoroughly. At this stage you should be able to see any small bones you may have missed.

Adjust the seasoning and add a little more lemon juice if needed.

Note:

This pate is best prepared a day ahead of time, and the flavours will develop over a couple of days. When it's first made, you'll hardly notice the garlic, but raw garlic can become quite powerful, so use it sparingly. Serve generously on rounds of freshly cut French bread, toasted if you wish and garnished with sprigs of dill.

Mackerel with Gooseberry Salsa and Crushed Potatoes

Serves 2

This recipe screams summer. At the same time the mackerel move inshore and start to get caught on the piers and harbour walls of Yorkshire seaside towns, the gooseberry bushes in our gardens are filled with ripe, zesty fruit. It's almost like the two know they need to be together! This recipe softens the sharp edges of the gooseberries with sugar and apple balsamic vinegar, complementing the mackerel beautifully.

Ingredients

Gooseberry salsa:

180g gooseberries, some sliced in two, some roughly chopped
4 tbsp caster sugar
3 tbsp apple balsamic vinegar (preferably Willy's)
Juice and zest of one lime
2 shallots, halved and finely sliced
2 tbsp mint, finely chopped
2 tbsp chives, finely chopped
Sea salt
Black pepper

Method

Whisk everything together and allow to stand for at least half an hour. It's best mixed together a few hours before you cook the mackerel.

Ingredients

Crushed Potatoes:

500g Waxy new potatoes
Sea salt
Black pepper
70g Jersey butter
2 tbsp flat leaf parsley, finely chopped

Method

Boil potatoes for 20 mins.

Crush, but don't mash, so you have some biggish pieces of potato still left.

Add butter, sea salt, black pepper and parsley. let the butter melt then fold it in gently, so as not to break up the potatoes too much. Allow to stand for a couple of minutes so the butter soaks into the potatoes before serving.

Ingredients

Mackerel:

2 tbsp sunflower oil
4 mackerel fillets
Sea salt
Black pepper

Method

Season the mackerel with salt and black pepper.

Heat the oil in a pan to medium/high temperature. Cook skin side first for 1-2 mins, depending on the size of the fillet, holding the fish down at first to prevent it curling up. Turn it over and cook for another 1-2 mins until cooked through.

Plate up the potatoes first, top with the mackerel, then spoon over the dressing. It will soak into those lovely buttery potatoes.

Serve immediately with a fresh green salad.

It's one of my favourite seasonal recipes and makes a really delicious plateful of food!

Mackerel with Ginger and Asian Slaw

Serves 2

A fishing trip to Whitby with our friend and former Great British Menu contestant, Tim Bilton, inspired this dish. Tim cooked freshly-caught mackerel, that we'd hauled straight from the sea aboard our boat *The Farmer's Girl* on strings of bright coloured feathers. He made a lovely, but simple, dish that combined shredded ginger, soy sauce, lime juice and zest. A quick, memorable and delicious feast, I've added an Asian slaw giving some texture and crunch alongside the flavours of the East. There is no finer fish than the humble mackerel when it's still bright and fresh from the sea.

Top Tip: Get everything ready before you start as this is very quick to cook.

Ingredients

2 tbsp oil
4 mackerel fillets
4 tbsp light soy sauce
2 tsp fish sauce
Thumb-sized piece of fresh ginger, finely julienned
Pickled ginger to garnish
Juice and zest of two limes
Bunch of spring onions, finely chopped

Recipe continues overleaf

Method

Add the oil to a preheated pan on a medium/high heat.

Place the mackerel skin side down in the pan, holding it down initially to prevent it curling. The fresher the mackerel, the more it will curl as you cook it.

Add the soy sauce and fish sauce – throw in the ginger, too, and cook for 1-2 minutes depending on the thickness of the fillets.

Turn the fish over and cook for 1-2 more minutes more. If the mackerel skin hasn't blistered enough, you could finish it off with a kitchen blowtorch.

Squeeze on the lime juice, and garnish with pickled ginger, lime zest and spring onions.

Ingredients

Asian Slaw:

Dressing

3 tbsp ginger wine vinegar (I use Slow Vinegar Company,
    you can use rice wine vinegar if you can't get the ginger)
3 tbsp soy sauce
2 tbsp fish sauce
Juice of one lime
3 tbsp groundnut oil
1 tbsp honey
2 cloves of garlic, crushed
Thumb-sized piece of ginger, finely chopped
2 tbsp gochujang
1 red chilli, finely diced
Handful chopped coriander

Slaw

You'll need a mixture of raw vegetables of varying colours, tastes and textures.

I suggest the following but choose what you enjoy:

250g white cabbage finely sliced
250g red cabbage finely sliced
2 carrots julienned
125g celeriac, peeled and grated
1 Red pepper sliced
150g edamame beans
Bunch of spring onions, finely sliced diagonally
2 tbsp toasted sesame seeds

Method

Whisk up the dressing ingredients and taste, add a little extra lime if you think it needs more. Allow to stand for at least half an hour, preferably several hours.

Whisk the dressing again before adding to the vegetables and sprinkle with the sesame seeds immediately before serving.

1977 – The Queen's Silver Jubilee

In July 1977, the year of the Queen's Silver Jubilee, The Queen and The Duke of Edinburgh visited Cannon Hall Museum, next door to the farm. We were pupils at Cawthorne School at the time and were due to be on stage when The Queen was there, doing our little bit of theatre. It was all about piped water coming to Cawthorne. It was drilled into us on so many occasions by our teachers, that every pupil I know from that time still remembers some of the text: 'If you're standing there, wondering why we're standing here, we'll tell you!' We were rehearsing it for months, time after time! Mrs Thorpe, our teacher, was determined we had to get it right for royalty!

I remember us being allowed to stay up late that night, dancing to a steel band at a party in the park in the evening after her visit, taking turns to limbo dance as darkness fell.

It was 12 years after The Queen's visit that we opened the farm up to the public. Since then, we've been fortunate enough to welcome two of the late Queen's children to the farm, Prince Edward, The Earl of Wessex, now The Duke of Edinburgh and, more recently, Anne, The Princess Royal, who opened our Lucky Pup Dog Cafe. It was such an honour to have royalty visit the farm and to join them for lunch.

Roast Rump of Lamb with Salsa Verde

Serves 2

We're proud of the excellent lamb that we produce at Cannon Hall Farm. Over his 82 years my dad, Roger, has presided over the birth of many thousands of lambs. My favourite roasting joint is lamb rump. It's a great way to produce a quick roast dinner midweek when you don't have time to cook a larger joint. It's a beautiful, tender joint that's already the perfect portion size for one. Choose those with a reasonable layer of fat as this will help to naturally baste the meat during cooking to keep it moist and succulent.

Ingredients

Olive oil
2 rumps of lamb, around 350g each
A few sprigs of rosemary, broken into smaller pieces
2 garlic cloves, sliced thinly
Salt and freshly ground black pepper

Recipe continues overleaf

Method

Rub a little olive oil on the fat that covers the lamb rump. I do this so that the salt and black pepper stick to the joint. Using a sharp knife, make a series of incisions in the fat on top of the lamb, and insert slices of garlic and sprigs of fresh rosemary into the cuts. Season with salt and freshly ground black pepper.

Place any leftover garlic or rosemary on a baking tray and put the joints on top. Cook at 200°C for 25-30 minutes.

Wrap the lamb in foil to keep it warm and rest it for 10 minutes.

Ingredients

Salsa Verde:

2 green heritage tomatoes, de-seeded and finely chopped
1 clove of garlic, crushed
2 tbsp wild garlic leaves, finely chopped (if available)
2 tbsp fresh mint leaves, finely chopped
2 tbsp fresh flat leaf parsley leaves, finely chopped
1 green chilli, de-seeded and finely chopped
Zest and juice of half an unwaxed lemon
1 shallot, finely chopped
6 tbsp extra virgin olive oil
1 tbsp Dijon mustard
8 anchovy fillets, finely chopped

Method

Combine all the ingredients in a bowl and mix well.

To make a quick version, you can roughly chop the items and use a hand blender. Cut the lamb into slices against the grain, and serve with a generous helping of the salsa verde, a green salad and some boiled new potatoes.

Note

You don't have to be too precise about quantities or ingredients. If wild garlic isn't available just leave it out or get creative and think of something else to include. It's important to note that 'salsa verde' simply means 'green sauce' – so ingredients can and do vary, depending on season and availability. These can include tomatillos, coriander, basil, spring onions, green pepper, capers, gherkins, lime juice or various vinegars as well as, or instead of, other ingredients. It is particularly important to taste this and season, or perhaps add more of the acidic element, in this case lemon juice.

'Your Father is Foul!'

There was a time back in the 1970s when the concept of a barbecue was a new and exciting one. The idea of grilling meat outside was an alien concept to us Brits back in those days. Mum, not wishing to be left behind, had purchased one of these new-fangled devices, and on one hot summer evening had determined that we were going to have a go at barbecueing.

Dad wasn't terribly keen on the idea, or about spending his hard-earned brass on a piece of metal of questionable value. For whatever reason, he'd taken against the idea of a barbecue and decided he was going to be grumpy. On occasions like this, Mum would declare Dad to be 'foul' (pronounced 'farl') which indicated to everyone that he was in a very bad mood indeed!

There are some who consider us to be the perfect family, but this was one of the many times where that has been far from the truth. With the pressures of work, kids and money at that time, there were plenty of reasons to be negative, and that day was certainly one of them.

No matter what he tried, Dad couldn't get the dratted thing to light. It would smoke for a few minutes, look like it was on the way, then die gently, along with our hopes of any sustenance. I remember my brothers and I wandering around the garden, feeling the negative vibes bouncing through the open window between my mum preparing salad stuff at the sink in the kitchen, and Dad's fruitless endeavours in the garden. I think they ended up cooking the meat in the oven. I'm not sure that barbecue ever saw the light of day again, in fact there's a good chance it's still in the corner of a barn somewhere at the farm.

Despite this early failure, the barbecue eventually became one of our favourite ways of cooking at the farm, and Dad has since enjoyed many a feast of farm shop burgers, sausages and marinated meat, cooked over the coals in the garden. Here's one of our favourite barbecue dishes.

Mill Farm BBQ Chicken with Tzatziki

Serves 4

Ingredients

8 bone-in chicken thighs
4 or 5 cloves of garlic, peeled and sliced
½ tsp sea salt
½ tsp coarsely ground black pepper
4 tbsp olive oil
1 unwaxed lemon, zested and thinly sliced
Small handful of fresh dried thyme sprigs,
    leaves stripped from stems
Small handful fresh dried rosemary sprigs,
    leaves stripped from stems
2 tsp chilli flakes

Recipe continues overleaf

Method

Make parallel cuts in the skin and flesh of the chicken pieces, to allow the marinade to be absorbed.

In a bowl, mix the garlic, sea salt, black pepper, olive oil, lemon zest and slices, thyme, rosemary and chilli flakes. Stir or preferably massage the marinade into the chicken pieces in the bowl. Make sure you wash your hands well after handling the raw chicken.

Leave to marinate in the fridge for a minimum of 4 hours, ideally longer. Give the chicken a stir every couple of hours to make sure the marinade gets absorbed. Don't worry if the garlic appears to turn blue, it's just an enzyme reaction with the lemon and is harmless.

Take the chicken out of the fridge an hour before cooking, and light the barbecue at least 30 minutes before starting to cook. I always use charcoal to cook outdoors these days. Gas just doesn't cut it for me. I think cooking outdoors is something of a primeval thing, and there's nothing primordial about a tank of gas and an artificial flame. Make sure the flames have subsided, and the embers are glowing white before you start barbecueing. Cooking should take roughly 25 minutes, but will depend on the size of chicken pieces. A meat thermometer is helpful when cooking chicken on the barbeque.

Ingredients

Tzatziki:

Half a large cucumber
Sea salt
450g Greek full fat yoghurt
1 clove of garlic, crushed
2 tbsp fresh lemon juice
1 tbsp of chopped fresh dill
2 tbsp chopped fresh mint leaves
1½ tbsp extra virgin olive oil
Ground black pepper

Method

Peel the cucumber and slice it in half lengthwise. Scrape out the seeds using a teaspoon and grate the cucumber flesh on to a clean tea towel to absorb the water. Sprinkle the grated cucumber with salt, and let it stand for 20 minutes. Squeeze the cucumber with the towel to remove any excess moisture.

Combine the Greek yoghurt, grated cucumber, crushed garlic, lemon juice, chopped dill, mint and one tablespoon of the olive oil and stir well. It's best to adjust the seasoning after a few hours as the flavours will become more intense.

Drizzle the remaining olive oil on top of the tzatziki and sprinkle with a few dill fronds to serve.

Note

Make the tzatziki a few hours ahead, so the flavours have time to develop. It can even be made the day before it's required.

BBQ Lamb Kebabs with Katchumber and Yoghurt Flatbreads

Serves 2

Men and fire; there's no mistaking the attraction. Fire up the barbecue, invite a few friends round and you can guarantee the blokes will all end up standing around the place where the cooking is getting done. They'll even be asking if they can help – I've known them to be almost fighting over the tongs and, of course, everyone is an expert in the art of cooking with fire! Here is one of my favourite barbecue dishes. I'll leave you to fight it out round the barbecue to decide who gets to cook it!

Ingredients

500g lamb from the leg or shoulder, cut into 5cm chunks
You will need: 4 wooden skewers

Marinade:

2 tsp sunflower oil
75g full fat Greek yoghurt
Thumb-size piece of ginger, peeled and finely grated
4 cloves of garlic, crushed
½ tsp fine sea salt
1 tsp chilli powder
Juice of one lime
2 tsp ground cumin
1 tsp coriander

Method

Mix the ingredients together and marinate the lamb pieces in it for at least 6 hours – preferably overnight.

Soak the skewers in water for several hours. This will help to stop them burning on the barbecue. Thread the lamb onto the skewers keeping as much of the marinade on them as possible and cook them on the barbecue, turning regularly for around 10 minutes altogether. Some charring on the meat is desirable.

Katchumber:

This looks more attractive thinly sliced rather than finely chopped.

Ingredients

½ red onion, thinly sliced
½ cucumber, sliced in half longways, seeds removed
      using a teaspoon, then thinly sliced
2 large tomatoes, de-seeded then sliced
½ firm mango, stoned and thinly sliced
Small bunch coriander, roughly chopped
      with a few leaves left whole
½ tsp fine sea salt
Juice of half a lime
Lime wedges

Method

Gently mix the onion, cucumber, tomatoes, mango and chopped coriander in a serving bowl. It's best to use your hands rather than a spoon, so the pieces don't break up.

Sprinkle with the salt and lime and mix again, then garnish with the coriander leaves and wedges of lime.

Recipe continues overleaf

Flatbreads:

These flatbreads are quick to make as you don't need to use yeast.

Ingredients

500g self-raising flour
300g whole Greek yoghurt
½ tsp salt
1 tsp baking powder
50g melted butter
Coriander leaves, roughly chopped

Method

Combine the flour, yoghurt, salt and baking powder together in a bowl, then knead the dough on a well-floured surface until it is no longer wet. Add a little more flour if the mixture is too sticky.

Wrap the dough in cling film and rest for half an hour.

Separate the dough into 4 equal-sized pieces and roll out thinly on a floured surface.

Cook in a dry frying pan on a medium heat for 2-3 minutes on each side, until the flatbreads have brown bubbles on them.

Brush with the melted butter and sprinkle with coriander.

Note

I sometimes add a drizzle of Lingham's Chilli Sauce to this if I'm in the mood for a little more spice. This dish also goes down a treat with my homemade lemonade.

Carnival Time

Cawthorne's Summer Carnival was the highlight of the year. Something that all the children of the village looked forward to. It was a celebration of village life, a real community event and everyone would get involved.

Each street in the village would produce a float. St Julien's Way was on the new estate in the middle of the village and they always chucked lots of money at theirs, producing some huge and impressive creations over the years. A pirate ship that almost took out the village telephones with its towering masts was one notable piece of work. We liked to think we went for quality rather than quantity, but it must be said, quantity almost always won! Being the competitive lot that we are, there were times we were a little bitter about our lack of success!

Our 'Chitty Chitty Bang Bang' was a cracking effort, but came nowhere! The car's wings even flapped as we drove along! Robert was dressed up as the Child Catcher with a large net and a prosthetic nose that kept falling off. David was one of his captured children in a large cage on the back of a wagon, and I was Grandad, complete with moustache. I spent the day parading round the village singing 'The posh, posh travelling life!' I always took my roles very seriously!

As I got into my teens, my artistic talents were put to work on various floats. I produced what I thought was a rather good 10ft-long cow jumping over the moon for the nursery rhymes float, and a mural of a time challenged rabbit, for 'Alice in Wonderland'.

In my memory, the sun always seemed to shine on carnival day and our float should always have won. We were delighted when we did eventually win with our float 'The Royal Mail' which was my dad's tractor and trailer cleverly disguised as a mail coach and horses – the horses brilliantly painted by my cousin Richard, who also inherited the artistic gene in the family. We were so proud of it we took it to The Mayor's Parade in Barnsley, and it won a prize there, too!

Raspberry and Mascarpone Crème Brûlée

My favourite dessert – bar none! This has everything I need at the end of a meal. Memorable for its sweet, crunchy, sugar topping, smooth, creamy custard and the sharpness of the raspberries at the bottom of the dish that cuts through all that sweetness.

Ingredients

120g raspberries
½ vanilla pod
300ml whipping cream
60g mascarpone
30g caster sugar, plus extra for topping
4 egg yolks

Recipe continues overleaf

Method

Share the raspberries between 4 individual ramekin dishes.

Split the vanilla pod and scrape out the seeds.

Boil the cream, mascarpone and vanilla pod and seeds together, then allow to cool until just warm.

Mix the sugar and eggs together and add to the cream mixture.

Return to a low heat and whisk until it becomes like a thick custard.

Strain through a sieve onto the raspberries and chill for 6 hours.

Sprinkle on a thin layer of caster sugar, before gently caramelising with a blowtorch. Be subtle, as it's easy to burn the sugar and make it bitter if you try to do it too fast. You can use a grill, but a decent blowtorch will deliver better results.

Put them back in the fridge for 1 hour before serving.

My First Food Memories

Probably the first food I remember eating and enjoying was boiled egg with soldiers, thin slices of toast cut into strips for the purpose of dipping. I've loved soft egg yolks ever since and always hated hard boiled ones. 'Clean as a whistle!' I would chirrup, holding up the empty shell for inspection every single time. I guess we all have our family rituals, and this was one of ours.

I remember being bewitched by the Angel Delight adverts on TV, but being disgusted by the stuff when I tried it. Even then I, like many others, was easily influenced by an effective marketing message.

I can still remember the day that put me off peas for life. Mum and Dad must have been busy because they'd asked Auntie Shirley to look after me. I remember being sat on her front doorstep on a hot summer's day, dressed in shorts, my knees grubby as always, eating cold, wrinkled peas that she had given me to keep me quiet. So grim were these peas that I've tried studiously to avoid peas ever since. I suspect I will never deliberately cook with them, but if you think they'll improve one of my recipes for you – I know many would add them to a fish pie – then feel free to include them.

AUTUMN

Although I love springtime, autumn runs it very close for my favourite season of the year. Keats so aptly titled it a season of mists and mellow fruitfulness, and as the days get shorter, I stock up the woodpile at Mill Farm to ensure that the fire will burn brightly in the hearth on the long winter evenings. It's a time for the farmer to gather in his harvest, the gardener to preserve the fruits of his labour, and the forager to take advantage of the free food that nature provides in abundance in the hedgerows, fields and woods.

Although I was little more than a toddler, I still can remember setting off mushroom hunting with Gran and Auntie Flo. Back in those days, what is now our event field at Cannon Hall Farm was excellent for growing wild mushrooms. Every autumn, as the cooler, wetter weather arrived, it would produce a crop of delicious, large brown-gilled field mushrooms. Mum would cook and freeze any that didn't get used. I remember them as so much tastier than the bland things we find in the shops now.

After the mushrooms had gone, Gran would keep us busy collecting sticks that had fallen from the trees to use as kindling for the fire. Nothing much was wasted back in those days.

Love in The Gift Shop

We opened the first gift shop on site in the early 1990s. We didn't have a huge range of produce in the beginning. Mum had knitted some jumpers made from the wool of our angora rabbits. And there were his and hers cow and bull mugs, decorated with cartoons I'd drawn. A fairly conventional looking cow with a modest udder was the 'hers' mug. A laconic looking bull with huge testicles on the 'his' version. That was our first venture into farm merch. We took on a shop assistant, as I had to muck the pigs out, clean the toilets and feed the rabbits before I could get into the shop.

One day myself and Anne, the pretty local lass that we'd hired – and a girl I'd long held a candle for – were dusting some of our rather limited product range. Wooden bowls and fruit, beautifully crafted by an elderly local woodturner. He'd lovingly labelled each item with a tag that showed it had been made from a local hardwood, oak, ash, alder or elm. As she polished a particularly handsome pear, Anne suddenly turned to me with her sparkling green eyes, and said, 'you know, I really love you!'

I was astonished, rather delighted, but quite taken aback. I looked at her, and following a long pause I managed to gasp, 'Do you? Really?'

She looked a little dreamlike, lost in her own thoughts as she ran a delicate finger across the smooth surface of the wood.

'Oh yes,' she replied, looking at me intensely, pausing then holding up the pear, with the handwritten tag, and saying, 'You know? The wood!'

I'm not always that quick on the uptake, but it rapidly dawned on me that when she'd said 'you' she'd actually meant 'yew'.

"Ah… yes… lovely." I replied as I deflated rapidly. I consoled myself that at least I hadn't declared 'I love you, too!'

I did have a couple of dates with Anne. But never heard those words again…

Anne is mostly vegetarian, so here I'm sharing one of my rare vegetarian recipes showcasing some of my favourite autumn ingredients.

Almost a Waldorf

Serves 2 as a main, 4 as a starter

Ingredients

1 crisp eating apple, cored and thinly sliced
Fennel bulb, thinly sliced, fronds reserved
80g mixed salad leaves
100g walnuts, roughly chopped
125g Yorkshire blue cheese, crumbled

Dressing

2 tbsp olive oil
2 tbsp apple balsamic vinegar
1 tsp runny honey
1 tbsp wholegrain mustard
1 tbsp Dijon mustard
1 shallot, finely chopped
1 tsp finely chopped thyme
Pinch of sea salt

Method

Whisk together all the dressing ingredients and add the apple and fennel slices. Stir and stand for 20 minutes.

Create a bed of salad leaves on the base of a serving plate, spoon the dressed apple and fennel over the top.

Sprinkle over the walnuts and blue cheese, drizzle with any remaining dressing and serve.

Quesadillas with Prawns and Squeaky Cheese

Serves 2

Here I combine some of the delicious ingredients that I've learned to love on my trips to the Yucatan Peninsula in Mexico with a halloumi style cheese that is made by a Syrian refugee here in Yorkshire. They come together with perfect synergy to make these delicious quesadillas.

Ingredients

1 tsp salt
1 tsp garlic powder
½ tsp onion powder
½ tsp chilli powder
¼ tsp black pepper
¼ tsp smoked paprika
½ tsp ground cumin
300g raw prawns
2 tbsp olive oil
25g unsalted butter
½ white onion, finely chopped
½ red pepper, finely diced
2 large tortillas
2 spring onions, chopped
220g pack Yorkshire Dama Squeaky Cheese with Chilli, grated

Method

Mix together the salt, garlic powder, onion powder, chilli powder, black pepper, smoked paprika and ground cumin.

Mix with the prawns and leave to marinate for half an hour.

Add the olive oil to a large frying pan on a medium/low heat and fry the white onion and pepper gently for about 5 minutes or until soft.

Add the prawns to the pan and cook for about 3 minutes until pink. Be careful not to overcook them or the prawns will be rubbery.

Remove from the pan, wipe the pan clean and melt the butter on a medium heat.

Place a tortilla on a plate and cover half of it with the prawn, onion and pepper mixture, then sprinkle over the spring onion and the grated cheese.

Fold the tortilla over and slide it into the pan with the melted butter. Cook for a couple of minutes on each side or until it's beginning to brown.

Cut into triangles and serve with sour cream, pico de gallo (see page 112), guacamole, sikil pak (see page 113), salsa or your favourite hot sauce.

Pico De Gallo

This is basically a chunky salsa — great as a dip with nachos, alongside scrambled eggs for a little kick at breakfast, or on the side of the quesadillas.

Ingredients

Tomatoes, skinned, seeded and diced
½ white onion, finely chopped
1 medium jalapeno, finely chopped
¼ cup lime juice
Small bunch coriander including the stalks, chopped

Method

Skin the tomatoes using a kitchen blowtorch or a dip into a jug of boiled water for a minute. Mix the tomatoes together with all of the other ingredients.

Note

This is best made a couple of hours in advance to allow the flavours to combine a little and is best served at room temperature.

Sikil Pak

Our Pumpkin Festival at the end of October is one of the biggest events of the year. What used to be one of the quieter school holidays has become one of the most successful after we launched this fun event. The children get dressed up in Halloween-themed costumes and the best fancy dress each day wins a prize.

The children take home a pumpkin lantern at the end of the day. Of course, pumpkins are for eating as well as carving and I discovered a great new use for pumpkin seeds on my last trip to Mexico.

Most people have tasted Mexican guacamole and the more conventional salsa recipes, but here's a different dip to try. I don't think it's particularly well known in the UK, so I thought I'd share it with you. This salsa is a traditional Mayan dish from the Yucatan Peninsula. It's a type of salsa made primarily with roasted tomatoes, pumpkin seeds and spices. It is served with tortilla chips or used as a topping for meats, tacos, and other dishes. It has a rich earthy flavour, balanced with the tanginess of the chilli and lime.

Ingredients

2 tomatoes, chopped into wedges and de-seeded
1 small red onion, chopped into wedges
2 cloves of garlic, left unpeeled
1 tbsp olive oil
100g roasted pumpkin seeds
½-1 red chilli pepper, roughly chopped
Juice of 1 lime
Small bunch of coriander, roughly chopped
Wensleydale cheese *optional*

Recipe continues overleaf

Method

Preheat the oven to 200°C.

Put the tomatoes, onion and garlic in a bowl, add the oil and coat before placing on a baking tray and roasting for 15 minutes.

Squeeze out the roasted garlic and add to a bowl with the onion and tomato, along with the pumpkin seeds, red chilli, lime and coriander.

Use a hand blender to create a thick paste. Use a little water to loosen up the mixture to the required consistency if necessary.

Everything needs a little bit of Yorkshire in it, and on this occasion, I tried crumbling Wensleydale cheese over the dip and its mildly acidic taste worked beautifully with the rest of the ingredients!

Creamy Tarragon Chicken

Serves 4

I love chicken thighs! For me they are the tastiest part of the chicken, they're affordable, incredibly adaptable and don't dry out easily during cooking like chicken breast can. This is so tasty, it's one of my favourite supper dishes. The tangy mustard and cream combines with the subtle aniseed taste of tarragon in a sauce that's guaranteed to get your tastebuds tingling.

Ingredients

8 bone-in, skin on chicken thighs
Sea salt and black pepper
2 tbsp olive oil
150g smoked streaky bacon, diced
200g chestnut mushrooms, halved
Half a large onion, finely chopped
1 heaped tbsp of chopped fresh tarragon
300ml chicken stock
250ml dry white wine
2 tbsp of wholegrain Mustard
1 tbsp Dijon Mustard
175ml double cream

Recipe continues overleaf

Method

Pre-heat the oven to 190°C.

Season the chicken thighs with sea salt and freshly ground black pepper.

Take a large, deep sauté pan and heat 1 tbsp of oil to a medium/high heat. Add the chicken to the pan skin down and brown for about 5 minutes, turning halfway through the cooking time.

Place the browned chicken on a baking tray and cook in the oven for 30 minutes.

Turn the heat down to medium and in the same pan add the rest of the oil and fry the bacon for 3-4 minutes until it starts to brown.

Add the diced onion and cook, stirring occasionally, until the onion is soft and translucent – usually about 2-3 minutes. Add the white wine, bring it to a simmer, and reduce by half until the liquid becomes syrupy.

Add the chopped mushrooms and stir in for 1 minute, then stir in the chicken stock and tarragon. Add both mustards to the pan, stir and simmer for about 5 minutes, reducing the heat slightly.

Add the cream and continue to simmer for about 10 minutes, stirring occasionally. Taste the sauce and adjust the seasoning.

Remove the chicken thighs from the oven and check they're cooked through. Here that meat thermometer comes in handy again – we're looking for 74°C. Add the thighs and juices from the chicken to the pan and stir until the thighs are coated.

Serve along with rice or potatoes and your favourite vegetables, with the sauce spooned over the thighs.

The Last Step

When I was young I would creep downstairs and sit on the bottom step when Mum and Dad had friends around, for what was a very rare dinner party.

Their guests would be Rosemary and Nigel, and Margaret and Ron, and it was always a celebration of one of their anniversaries. I'd listen to the gentle hum of conversation and the laughter as they served up 70s food from Mum's Marks & Spencer cookbook. I wasn't really interested in what they were talking about, or the food at that time. I just wanted to feel a little closer to them. I'd sit there, rest my head against the third stair, curl up and fall asleep.

I'd wake next morning in my bed, having been discovered there, sleeping on the stairs, and carried up to bed.

I'm sure a classic fish pie would have gone down beautifully at one of those dinner parties.

Fish Pie

Serves 4

A good fish pie is comfort food at its very best. Rich and unctuous, this pie combines haddock, smoked haddock, king prawns and spinach to make a very special fish supper. A crisp mashed potato crust hides a creamy, smoky cheese sauce and tender pieces of fish. Spinach is a great addition that adds taste and texture to the finished dish. Ring the changes with the fish if you wish, any member of the cod family is good, but this combination works so well. I'd always include the smoked haddock which gives the dish an extra dimension.

This is another recipe that takes a while and creates a lot of washing up, but trust me it's worth it! It's a great recipe to make in advance and then pop into the oven as your guests arrive. Make sure you pick traditionally smoked haddock, which is a pale brown colour rather than the bright yellow dyed-smoked haddock. It has an infinitely superior flavour. You can use large raw frozen king prawns. They don't take long to defrost in a bowl of cold water.

Ingredients

650g fresh haddock fillet
400g naturally smoked haddock fillet
1 large onion
20 cloves
1.1 litres whole milk
2 carrots, peeled and chopped in two
2 sticks of celery, chopped in two
12 black peppercorns
Bouquet garni
2 bay leaves
300g spinach, stalks removed
50g butter
50g plain flour
2 tbsp Dijon mustard
400g mature cheddar cheese
12 large raw king prawns
6 or 7 large potatoes, peeled and cut into 1.5in pieces
Knob of butter
Splash of milk
Sea salt and black pepper

Method

Preheat the oven to 200°C. Skin the fish and cut into large chunks. Peel the onion and stud with cloves. Place the milk, onion, carrot, celery, black peppercorns, bouquet garni and bay leaves into a stockpot with the fish. Bring to the boil and gently simmer for 20 mins. I always seem to boil it over, so keep a careful eye on it!

Strain through a sieve, reserving the milk. It is full of smoky flavour and makes a lovely base for your sauce. Try to prevent the fish from breaking up (we're after large flaky pieces). Discard the vegetables, peppercorns, bouquet garni and bay leaves.

Recipe continues overleaf

Wilt the spinach in a separate pan and squeeze out all of the excess liquid with a wooden spoon, leave to one side to drain.

Mix the butter and flour, stirring it for about 5 minutes on a medium/low heat to make a roux. Add the flavoured milk bit by bit while whisking. Just using a small amount at first to obtain a smooth sauce, then increase the amount of milk until it is completely incorporated into a thick sauce. Add the mustard and cheese and stir until the cheese melts.

Lay out the cooked fish across the base of an ovenproof dish. Dot the raw prawns and teaspoons of the spinach among the fish and pour the sauce over.

Meanwhile, boil the potatoes for 20 minutes or until soft. Put them through a potato ricer (or mash them), add the knob of butter, a splash of milk, salt and black pepper and beat until smooth. Add the butter and milk sparingly, as you don't want your mash too soft or it will sink into the pie.

Add the mash to the top of the pie and spread it out with a fork, roughing up the top to create a textured finish. You can sprinkle the mash with more cheese if you wish, but I reckon it's just fine without.

Cook in the oven for 30 mins if cooking from room temperature. Cook 40-45 if cooking from chilled and insert a metal skewer into the centre of the pie to check it's completely warmed through before serving.

Note

If I'm making the pie just before I serve it, I add the raw prawns and pour on the warm sauce, before popping it in the oven. When I make it ahead of time, I let the component parts cool before putting the pie together.

Farmer Robert's Roasted Duck Breast with Pomegranate Molasses and Stir-Fried Purple Sprouting Broccoli

Serves 2

My brother Robert loves a perfectly cooked duck breast, and I created this quick and simple recipe just for him!

Ingredients

2 duck breasts, fat side lightly scored with a criss-cross pattern
Sea salt and black pepper
2 tbsp olive oil
2 tbsp pomegranate molasses
Handful of Pomegranate seeds

Method

Preheat the oven to 200°C. Dry any moisture from the duck breast and season with salt and pepper.

Heat the oil to a medium/high temperature in a frying pan and cook the fat side of the duck for roughly 3 minutes, until it is browned, and then turn it over and brown the other side.

Place them on a baking tray and cook in the oven for 10 to 15 minutes depending how you like your duck cooked. Drizzle on the pomegranate molasses 5 minutes before the end of the cooking time. Keep it warm and rest it for 5 minutes before slicing. Pour over the juices from the baking tray and sprinkle with some of the pomegranate seeds to serve.

Stir Fried Broccoli

Ingredients

Knob of butter
Purple sprouting broccoli
1 tbsp sliced almonds
2 tsp capers
2 tbsp pomegranate molasses

Method

Drain off some of the duck fat from the frying pan, leaving a couple of tablespoons and add the knob of butter. Stir fry the broccoli on a medium/high heat quickly, while the duck breast is resting.

After about 3 minutes, sprinkle over the almonds and capers and drizzle the pomegranate molasses. Warm them through for another minute, then serve alongside the duck, sprinkled with remaining pomegranate seeds.

Roasted Beetroot and Red Onions on Whipped Feta

Serves 4 (or 1 hungry photographer)

This is another delicious vegetarian option. Earthy root vegetables served with a zesty, whipped feta dip. Trust me when I say the photographer made short work of this one after he'd shot it!

Ingredients

3 tbsp olive oil
2 beetroots, peeled and sliced into wedges
2 red onions, peeled but keeping the root intact,
    sliced into wedges from top to bottom
1 small purple sweet potato, cut into wedges
4 sprigs of thyme
4 sprigs of rosemary
2 tbsp balsamic vinegar
200g feta cheese
100g full fat Greek yoghurt
2 tbsp lemon juice
1 large clove of garlic, crushed
Sea salt and black pepper
2 tbsp honey
2 tbsp extra virgin olive oil
2 tsp toasted pine nuts
Dill or rosemary sprigs to garnish

Method

Preheat the oven to 190°C.

Heat the olive oil in a roasting tray, then add the beetroot, red onion, sweet potato, thyme, rosemary and balsamic vinegar. Give the ingredients a good mix to ensure all the vegetables are well coated with the oil and roast for 40 minutes, giving the roasting tray a shake every 15 minutes or so.

Meanwhile, combine the feta cheese, yoghurt, lemon juice and garlic in a bowl with a little ground black pepper and blend together using a stick blender. Only add salt after tasting, as the feta can be quite salty. Adjust the seasoning and add extra lemon juice if required, but bear in mind that the flavours will develop over time.

Coat the centre of a plate with the whipped feta, drizzle with the honey and extra virgin olive oil. Sprinkle with the toasted pine nuts, dot with the herbs and serve with the roasted vegetables.

Going Underground

One of the great delights of growing up next to the parkland of Cannon Hall is the history that can be discovered in the landscape. Recently, the old icehouse has been renovated and the public can now peer down into its depths, where ice that was gathered during the winter was kept and used to chill food later in the year.

Having this sort of history around you was great as a kid. When you're reading Enid Blyton stories at night, living in a landscape where such adventures could easily be imagined is wonderful, but it can get you into trouble too.

Underneath the field next to the farm was one such trouble spot! A brick-lined, underground water tank, fed with fresh water from a subterranean stream. It was covered by a big stone flag, which you could pull up and rest on its edge to allow you to climb down into the little cavern below. You could perch on the ledge below, just above the still pool of water with its curved brick roof. Without a torch and with only a square of light from above to illuminate the scene, it was impossible to tell how far the water reached. The first time I'd been down there, I'd imagined it might stretch away for hundreds of yards, to who knows where. We did later establish that it was sadly, only a few yards in length.

One day we liberated a trout from a local stream, carried it in a bucket to this underground lair and released it. It lived there quite happily for a year or two, no doubt surviving on insects and amphibians that fell through the cracks around the stone. We often visited our pet trout in this little cave as part of the day's adventures. It would swim from the depths when the light flooded into its cave to see if we had some food for it.

Once, I'd been fishing. I was always fishing. It must have been a slow day, and I ended up at the hole in the ground, alone. I decided I'd visit our pet trout, dumped my tackle on the grass, lifted the stone and slid down from the bright sunlight into the cool space beneath, alighting on the ledge next to the water. You could crouch there. You couldn't stand upright.

Maybe it was a breath of breeze, or the soil above me shifted slightly, anyhow, something must have moved the stone. It slammed shut above my head. Thankfully it didn't hit me. I could have easily been knocked out, or worse still, fallen into the water and drowned.

I was in the dark, crouched on the ledge. A little light seeped around the heavy stone flag above me, illuminating my companions down here in this godforsaken place. They were a couple of tiny, forlorn looking skeletons. A frog and a mouse that had fallen through the gap around the stone, found they had no way to escape and died a sad and lonely death down there.

I pushed at the stone. It moved a few inches, but not enough to open. I tried again and again, more than a little embarrassed about the situation I'd got myself in. I was trapped; I couldn't get out! No one knew where I was. I could die here!

So, I did what all resourceful young men do at times like these. I took a deep breath, opened my mouth and screamed as loudly as possible for my mum! This hole in the ground is several hundred yards away from the house. How long I yelled for I have no idea, but I screamed and shouted for my mum for all I was worth! It might have been 20 minutes, it might have been half an hour, but eventually the stone lifted, and Mum's face appeared. It was such a huge relief! I'd felt I could be stuck down there forever.

I was lucky, as she'd been outside for some reason, maybe hanging out clothes to dry. She could hear my muffled cries, but couldn't locate where the noise was coming from. She'd been running around the farmyard, trying to figure out where I was. In the end she came up into the field and spotted my fishing box and rod next to that miserable hole in the ground. How long it would have taken her to find me if it hadn't been for that, heaven only knows.

Date and Walnut Loaf with Blue Cheese Butter

This has been on the menu (minus the blue cheese butter) since the day we opened Home Farm Tearooms in 1981. It's a simple recipe that has always been popular.

Ingredients

170g self-raising flour
110g caster sugar
60g walnuts, roughly chopped
110g dates, chopped
110g margarine, melted
2 eggs, beaten

Method

Mix together the flour, sugar, walnuts and dates. Pour on the melted margarine and mix. Add the eggs and mix. Grease a loaf tin with margarine, add a little flour and coat the inside of the tin, discarding any excess.

Bake at 150°C for one hour. Use a skewer to make sure it's fully baked. Serve with butter.

Blue Cheese Butter

Serve this with the date and walnut loaf if you're feeling adventurous. I'd made it to go with steak, but a viewer on Facebook live suggested trying it with the loaf and I was surprised how well it worked. Happy accidents have been shaping cookery since mankind first heated a pot over a fire.

Ingredients

125g butter
125g Yorkshire blue or blue Stilton cheese
1 tbsp fresh rosemary, finely chopped
2 tsp fresh sage, finely chopped
1 tbsp fresh parsley, finely chopped
Sea salt and black pepper

Method

Soften the butter for a few seconds in the microwave or in a bowl over simmering water.

Chop the cheese into small pieces, add the herbs, a pinch of salt, and a few grinds of black pepper.

Mash it all together and spread onto the date and walnut loaf generously.

Note

Another great way of using this adaptable butter is to mould the butter into a tube shape, wrap it in cling film and chill in the fridge, ready to slice off and melt onto steaks and other meats during cooking.

Mum's Ginger Cake
(A bonfire night favourite)

This recipe is almost always doubled, as my mum says we'd have it all eaten in five minutes if it wasn't. This must be Mum's most baked family recipe, and all the kids and grandkids love it. We normally bake it in a large square baking tray, as we think the bits without edges are always the best. Great granddaughter Nelly loves it, too. It must be something in the Nicholson genes.

We did try selling this in Home Farm Tearooms when it first opened, but it never sold very well. Despite the fact it tastes delicious, we could never get it to look particularly spectacular on a plate in the counter.

Ingredients

240g self-raising flour
pinch of salt
85g of margarine
85g sugar
1 heaped tsp ground ginger
2 tbsp golden syrup
Not quite a full cup of warm, not boiling water
    (exactly as written in Mum's recipe book!)
1 small egg

Recipe continues overleaf

Method

Preheat the oven to 150°C.

Put the flour, salt and margarine in a bowl and rub in until you have a crumby texture. Add the sugar and ginger and stir in. Melt the golden syrup in a pan with the warm water until it dissolves. Warm, do not boil. Add this to the mixture and stir in. Add the egg at the end and whisk in.

Cook for 30 mins at 150°C. Check the cake is cooked by inserting a skewer into the centre of the cake. If it comes out clean it's cooked.

Note

At its best this is soft, moist and sticky on top. Mum says everyone's oven is different and when she's changed ovens over the years, she's cooked this at temperatures anywhere between 150°C and 180°C. Get to know your oven and don't follow recipes slavishly. If this dish (or indeed any other recipe) doesn't turn out how you expect, think it through, and change something a little. That, for me, is part of the adventure of cooking.

Auntie Beryl's Lemon Meringue Pie

During the writing of this book, on 17th December 2023, we sadly lost Auntie Beryl, Dad's youngest sister, at the age of 92. In homage to her baking prowess, her recipe for lemon meringue pie was printed on the order of service at her funeral, so as a tribute to her I'm also including it here.

Auntie Beryl and Uncle Brian ran a newsagent's shop in Snaith for many years. When we first opened our tearoom in 1981, she showed me the basics of using a till. She told me all the notes should be the same way up and facing in the same direction when they were put in the till, to make it easier when it came to cashing up.

Our till was an ancient brass monstrosity that was calibrated in pounds, shillings and pence and used sliders rather than buttons. The only parts of it that worked were the large clunky handle on the side and the cash drawer. I would crank it round a couple of times to open the drawer, which would fire out aggressively into my midriff. That till must have had the strongest springs in the world!

She was a real cheerleader for the farm, enthusiastically telling her family and friends about each new development and TV show when it came along. She always enjoyed her visits back to Barnsley to see Dad and Mum after her retirement in Scarborough. My cousin Anne told me Auntie Beryl would've loved the idea of having one of her recipes included.

You could make a quick version of this using a shop-bought pastry case, but if you want to go the whole hog, here's the pastry recipe.

I believe it's a recipe that comes from a thriftier post-war age, so it may be a little light on some of the more expensive ingredients and while it's delicious, it is quite understated for modern tastes, so when I cooked it, I added the juice and zest of two lemons instead of one. I also added an extra egg and replaced 50ml of the milk with double cream, to make it a little more luxurious. That's the way recipes evolve! You choose where you'd like to go with it! Both versions are very tasty.

Ingredients

Pastry:

170g plain flour
Pinch of salt
25g icing sugar
100g cold butter, cut into small cubes
1 egg yolk
1-2 tsp cold water

Method

Preheat the oven to 150°C. Place the flour, salt and icing sugar in a mixing bowl, stir to combine. Rub in the butter until the mixture resembles breadcrumbs. Add the egg yolk and 1 tsp of water and mix until the dough comes together. If there are still loose crumbs add a little extra water.

Wrap in cling film and leave it in the fridge for at least 30 minutes. It's important to handle the pastry as little as possible.

Roll the pastry out and use it to line a 20cm tart case. Remove any spare pastry around the edges with a rolling pin or sharp knife. Add a layer of baking parchment, weigh it down with baking beans or pulses and bake for 10-15 minutes, until the pastry is firm to the touch and just beginning to colour lightly. Remove the beans and return to the oven for a further 5 minutes.

Ingredients

Filling

275ml full fat milk (or 225ml + 50ml double cream)
3 heaped tsp cornflour
60g caster sugar
2 egg yolks (or 3 egg yolks, see intro)
Zest and juice of 1 lemon (or 2 lemons for fuller flavour)
Knob of butter

Method

Boil the milk and mix in the cornflour.

In a separate bowl, mix the egg yolks and sugar together. Allow the milk to cool somewhat then return to the heat as you whisk in the egg and sugar mixture. Add the juice, zest and the knob of butter and whisk until the sauce thickens to a thick custard.

Pour into the pastry case being careful not to overfill, then when cooled slightly, top with the meringue.

Recipe continues overleaf

Ingredients

Meringue topping:

3 egg whites
180g caster sugar

Method

Whisk the egg whites, slowly at first, then building up speed, adding the sugar a tablespoon at a time until half the sugar has been added and it forms into stiff peaks.

Fold the remaining sugar carefully into the mixture, so as not to lose too much of the air. Spoon or pipe the meringue onto the pie, creating small peaks.

Place the pie in the oven and cook for between 15 and 20 minutes. Check the pie – the meringue should be starting to brown but not be burning. When the pie is browned and has a crisp shell remove it from the oven. Cool for 30 minutes and refrigerate uncovered.

Last of The Autumn Wine

In the early years our ever-industrious mum, Cynthia, got into homemade wine making. The airing cupboard would be filled, not just with neatly folded clothes, but large glass demijohns, slowly bubbling away as the fruit turned into alcohol. She produced many a classic vintage using plums, crab-apples, elderberries, peaches and all manner of different fruits. Mum liked making it but didn't drink much of it. Dad, on the other hand, loved the rhubarb wine she made and enjoyed a particularly enthusiastic drinking session one Christmas.

'To be fair to him, he was never a big drinker,' Mum says, 'not really used to it. I woke the next morning to find him lying flat out on the living room floor. I left him there and went to Halifax with you three to visit your grandma and grandad. When I came back in the evening, he was still there – where I'd left him! I had to feed all the animals as well as you lot!'

Mum got so good at wine making she won a first prize at Penistone Show. Dad never was very good at drinking, perhaps that day put him off for good?

I remember setting off into Deffer Wood on an expedition with the family as a small boy, walking through bracken far taller than me, feeling like I might be on safari, in one of the Johnny Weissmuller *Tarzan* films that we loved. On a mission, not to the elephant's graveyard, but to collect crab apples to make into wine and jelly. We returned home with bulging carrier bags full of them. Getting children involved with collecting ingredients is a wonderful pastime and one I really do recommend to parents today, if you can detach them from their phones.

They were simpler, better times.

There's a real crisis in the mental health of children and adolescents today. One of the things we don't do as much is spend time with our children doing tasks, making them feel their contribution is valued. I think everyone needs to feel useful, from young children to the very old, they are much more likely to fulfil their potential if they feel appreciated and part of the bigger picture. Money and gadgets can't replace genuine parental or family interest and interaction.

WINTER

I wake with the first rays of sunshine on a winter's morning, and the light seems brighter than I might have expected. I open the curtains, and a surge of childlike excitement runs through me, as I look out on a field of freshly fallen snow. I open the front door and Riley pushes past me tearing around the garden in ever decreasing circles like a wild thing, his ears flapping and tongue lolling, until he comes to a screeching halt. Looking up at me with that daft grin on his face, snow crystals sparkle on his cream-coloured fur.

On days like these, we head down the valley, dragging our sledges with us, to the steepest field that runs down to the stream. It's a truly epic slope, and you really can go like a rocket on our old family heirloom – the 1930s American, flexible flyer sledge. Every generation of our family used it as children, and most of us as adults, too. Flying down that slope at whatever age you are gets the heart beating and fills your lungs with frosted air. You may take a fall at the bottom of the slope, rolling through the snow until you come to a breathless stop, eyebrows crusted with ice. When will I grow out of this? Never!

Then we tramp back to Mill Farm, our breath coming in great clouds as the chatter flows and the snowballs fly. All the family together. Stepping through the door, the warmth of the air in this little cottage hits us and warms us immediately. We pull off wellingtons and scarves and sit around the woodburner, hands clasped around steaming mugs of hot chocolate. We talk of winters past, of long-gone Christmas times, of sore fingers from plucking turkeys on a cold winter day. Of tramping through snow to rescue lambs in snowdrifts, and feed hay to the sheep. Times that were sometimes hard and unforgiving have gained a rosy glow, softened by the passage of time.

We sit and talk of family, and loved ones, and times long past, and in that warm room filled with laughter and memories, we remember, and smile.

Clare's Pasta Bake

Serves 4

My partner, Clare, makes a deliciously tasty pasta bake. It has a crunchy topping with a creamy sauce beneath, the spiciness of the pepperoni and a cheesy tang from the sauce make it the perfect comfort food. I always make sure I have pepperoni in the fridge just in case Clare volunteers to cook this. Sometimes it's fun to spice it up even more with some nduja sausage.

Ingredients

Sauce:

1 medium red onion, chopped
2 tbsp olive oil
4 garlic cloves, chopped
1 tsp dried oregano
650g passata
1 tsp sugar
Sea salt and black pepper
3 tbsp grated parmesan,
125g Mascarpone cheese

Method

Fry the chopped onion in the oil on a medium heat until soft and translucent, then add the garlic and oregano and fry for another minute.

Add the passata and warm through until simmering.

Add the sugar and seasoning. Simmer for another minute or so and if happy with taste, take off the heat and let it cool for a few minutes. Use a stick blender to make the mixture into a smooth sauce. Stir in the parmesan and mascarpone cheese just before adding to the pasta.

Ingredients

Pasta Bake

350g dried pasta – preferably rigatoni
Sea salt
5 rashers of smoked streaky bacon, diced
Tomato sauce (as above)
120g of sliced pepperoni
1 small red onion, finely chopped
2 garlic cloves, finely chopped
100g buffalo mozzarella, torn into small pieces
250g grated mature cheddar
1½ tsp chilli flakes *optional*
Grated parmesan and fresh basil leaves to garnish

Recipe continues overleaf

Method

Boil a large pan of salted water and cook the pasta for 2 minutes less than the recommended time.

Fry the bacon in olive oil until crisp and put to one side.

Add the chopped onion to the same pan and cook on a low to medium heat until soft and translucent, then add the garlic and cook for a further minute.

Add the pepperoni and fry for a minute more.

Add the bacon back to the pan and mix in the tomato sauce.

Simmer for 3 or 4 minutes then combine the sauce with the pasta.

Add half the torn mozzarella and 100g of the grated cheddar and stir.

Pour the mixture into an ovenproof dish and top with the other half of the torn mozzarella and the rest of the grated cheddar. Sprinkle the chilli flakes over the dish if required and bake for 25 minutes.

Serve with the parmesan and torn basil.

Nativity

My brothers and I went to the little village school in nearby Cawthorne. In most Church of England schools, each year, there is a nativity play. One year, I was chosen to be a pig! No teacher before or since, ever imagined there was a pig in that stable in Bethlehem, but Mrs Thorpe knew better!

Mum made me a mask and dyed a load of old light-coloured clothes pale pink, she was good at that sort of thing. I even had pink tights over my hands, so they'd look more like trotters. I was determined to take my part very seriously indeed!

'Oink, oink, I have brought some apples for the baby Jesus' …that was my line.

I oinked a lot after that, big authentic piggy snorts, to fill the time and establish myself as an important character in the well-loved tale. Trying to keep myself busy, I also spent my time on stage polishing that apple so much that it shone… after all, it was a special apple for the baby Jesus! I can remember some of the parents chuckling in the audience. I became so absorbed with my part, and attacked it with such enthusiasm that, from that day forward, all the other kids at school nicknamed me 'Oinky'.

I've tried hard to rise above it since, with limited success.

Since those early days I've been much happier cooking pork than I have been playing it. Here is Farmer Roger's recipe for perfect pork loin with apple sauce, and my recipe for the perfect pork crackling.

Farmer Roger's Pork Loin with Perfect Crackling, Roasted Root Vegetables and Apple Sauce

Serves 4

When it comes to creating great pork crackling, the biggest mistake people make is to choose too lean a piece of pork. You need a decent layer of fat – and that brings with it another advantage, the fat protects and bastes the meat as it cooks, keeping it moist. Lean pork with very little fat may appear to offer better value for money and offer a healthier option but will only lead to a dry, tasteless joint with poor quality crackling. If you like your food to taste great, don't be frightened off by a little fat, that's where the flavour and succulence comes from.

Ingredients

1.5 kg pork loin on the bone, scored and chined
1 tsp olive oil
Sea salt

Method

Preheat the oven to 225°C.

Place the joint into a shallow roasting tin. We use a shallow one to allow the heat of the oven to get to the skin. Dry the skin of the pork and rub with a little olive oil. This helps the salt to stick to the pork rind. Sprinkle the skin liberally with plenty of sea salt.

Place the pork, uncovered, on the highest shelf of the oven that it will fit on. Cook for 30 minutes then check to see if it has begun to crackle. If it hasn't, give it another 20 minutes and check it again. As soon as you're happy that the fat is browning and puffing up you can turn it down to 190°C and cook it for an additional 70 minutes to 1½ hours – to make around 2 hours in total.

After it's been in the oven for 1 hour 30 minutes, check the core temperature with a probe. You're looking for an internal temperature of 65°C. I'd normally expect it to take around 2 hours, but depending on the thickness of the joint, that can vary.

Wrap the pork in foil and let it rest for 30 minutes before carving, don't worry the crackling will remain crisp! This gives you plenty of time to make gravy.

Ingredients

Roasted Root Vegetables

3 tbsp olive oil
600g of mixed root vegetables, choose from parsnip,
    celeriac, carrot, beetroot or sweet potato
Handful of rosemary sprigs
Handful of thyme sprigs
Sea salt and black pepper

Method

Preheat the oven to 200°C.

Add the oil to a shallow roasting dish and warm in the oven for 10 minutes.

Peel and cut the vegetables into chunks, at angles. I prefer this to cubes as it means the thinner pieces receive more charring giving more taste and variety to each bite.

Place the vegetables and herbs in the roasting dish making sure they are coated with the oil, and season well. Cook for 30 mins, shaking the pan halfway through.

Ingredients

Apple Sauce

2 large Bramley apples, peeled and sliced
1-2 tbsp sugar

Method

Peel and slice the apples.

Add to a medium size pan with 200ml of water, simmer for 10-15 minutes, stirring regularly, until the apple has softened and broken up.

Add a tablespoon of sugar, stir and simmer for another minute. Taste and add more sugar if needed.

Serve with diced roasted potatoes in goose fat (see page 182) or roasted garlic mashed potatoes (see page 190).

Farmer Marshall's Sirloin Steak with Yorkshire Blue Cheese and Portobello Mushroom Sauce

Serves 2

My son Marshall has a limited cooking repertoire, but he loves a farm shop steak and makes a great job of cooking it! When choosing your steaks, you should look for darkish red, well-marbled meat. Lean, bright red supermarket steaks are best avoided, as they just don't hang them for long enough. Hanging beef properly costs money, both in refrigeration costs and in weight loss, and supermarkets aren't fond of anything that interferes with their profits. Steaks should have been dry aged for at least three weeks.

Ask your local butcher or farm shop how long they hang their beef for, that will tell you all you need to know about how much they value delivering a quality product.

Ingredients

Steak

2 sirloin steaks, well marbled and around
    1 inch thick with a decent fat covering
Sea salt and black pepper
Olive oil
A generous knob of butter
A few sprigs of thyme
3 garlic cloves, bruised

Sauce

A small glass of white wine
1 shallot, finely chopped
2 large portobello mushrooms, sliced
100ml chicken stock
100ml double cream
100g Yorkshire blue cheese

Method

Take the steaks out of the fridge an hour before you want to cook them, to allow them to reach room temperature, and season generously with sea salt and black pepper.

Add a glug of oil to a heavy-based frying pan. Get it really hot, so it begins to smoke. Cook the steaks for 1 minute on the fatty side-edge first, using tongs to hold it down and sear it. Then cook for 1 minute on each side.

Add the butter, thyme and garlic to the pan and cook for a further 1 minute on each side for rare, basting the meat with the butter and oil. Cook for 2 minutes extra for medium, and 3-4 minutes extra for well done.

Remove the steak from the pan and keep warm, allowing the meat to rest while you make the sauce.

Deglaze the pan with the white wine on a medium heat and reduce for a minute or two. Add the shallots and mushrooms and fry gently for a couple more minutes.

Add the chicken stock and reduce by half, before stirring in the cream.

Cook for a couple more minutes, to reduce the sauce still further, then stir in two thirds of the cheese and season to taste.

Crumble the rest of the cheese over the steak before pouring on the sauce.

Serve with twice cooked chips.

Note

Touch the steak firmly with your finger to check how well cooked it is. If it's soft it will be blue or rare. If it still gives a little, it's medium. If it's firm to the touch it will be well done. I don't recommend cooking it any more than medium rare, but you go ahead and do what you wish – it's your steak to ruin... but don't blame me!

Winter Sunday Evenings

Our parents were wonderful at indulging and involving us as children. We all have vivid memories of a particular period when we were growing up. Mum would cook us a special tea; she'd make a pizza dough and allow us to have whatever topping we wished. Bizarrely, mine was tinned sardines in oil with no tomato base! A strange, rather dry and ugly invention that was never going to make it into this cookbook!

After tea we'd play the board game Mousetrap. A game where you build an elaborate trap to catch a mouse. We loved it! Our evening would be completed by watching the TV adaptation of James Herriot's wonderful creation, *All Creatures Great and Small*, which was filmed in the Yorkshire Dales. That was a perfect Sunday for us, before the drudgery of school the next day.

They were wonderful nostalgic times that combined creating food with being competitive, something we Nicholsons have always embraced. I'm sure this Toad in t'Ole will go down rather better than one of those desperate pizzas I used to be so keen on!

Cumberland Ring Toad in t'Ole with Red Onion Gravy

Serves 2-4

This looks impressive, is inexpensive and can be whipped up pretty quickly. Light fluffy Yorkshire pudding combined with the finest of farm shop sausages and rich gravy makes for a really tasty meal when you don't have the time or the inclination to cook a proper roast with all the trimmings.

This makes enough for two good-sized puddings, with a bit left over for a few small ones.

Ingredients

140g plain flour
4 eggs
200ml whole fat milk
Salt and pepper
Sunflower oil
2 Cumberland ring sausages

Method

First make the batter. Mix the flour and eggs together in a bowl, then add the milk gradually while whisking to avoid lumps.

Season with salt and pepper.

Some suggest you rest the mixture for a few hours, but I tend to make it just half an hour in advance and haven't had any problems.

Preheat the oven to 225°C. I use heavy-based, round cake tins. Whatever container you choose should be thick walled, so it holds the heat as you pour in the batter. Use enough sunflower oil to coat the bottom of the tins, and heat the oil in the oven for at least 10 minutes.

Add the Cumberland sausage and cook for 5 minutes, then take the tin out of the oven, remove the sausage, pour in enough batter to cover the base of the tin and return the sausage to the middle of the tin, browned side up.

Cook on the top shelf of the oven for 23 minutes.

If you've done everything correctly, they will rise considerably, so make sure your oven rack isn't ultra close to the top of the oven or you'll be unsticking them from the roof!

Note
Do not open the oven door to check on your pudding before the end of the cooking time! Nothing but disaster will befall you if you do this!

Red Onion Gravy

Ingredients

Roasting juices from a joint of beef or
    pork, or two beef stock cubes
Large glass of red wine
2 red onions, sliced
1 brown onion, sliced
Few sprigs of thyme
Few sprigs of rosemary
1 bayleaf
500ml beef stock
Sea salt and black pepper
Cornflour or beef gravy granules (but don't admit it to anyone!)

Method

Tip away excess fat, but leave around 2 tbsp. Deglaze the roasting pan with the wine, scraping any baked-on residue off the base, stirring it into the wine. Reduce until you have quite a thick, syrupy consistency, then add the onions, thyme, rosemary and bayleaf.

Cook for 3 or 4 minutes until the onions are starting to soften a little, then crumble on the stock cubes and cook for another minute or so. They become jammy, melting onto the onions and sticking to them, intensifying the flavour.

Add the beef stock then bring to the boil and simmer for a few minutes to infuse the flavours from the herbs and reduce and thicken the liquid. If you need more volume, add more stock or water, preferably water that's been used for boiling vegetables, particularly potatoes.

Taste and season with salt and pepper.

If the gravy is already rich and full, thicken with some cornflour mixed with a little water, if it's thin and lacking taste you can cheat a bit by sprinkling in some gravy granules. You won't be the first or the last to do this – I won't tell if you don't!

Note

Gravy is always best made using the juices from a roasting joint, usually beef in this instance, so I advise you make this gravy after you've roasted some beef. You can serve some with the beef and keep the rest for a few days in the fridge, warming it through when you're ready to serve it up with your toad in 't 'ole. If you don't have meat juices, then start with some oil and the onions, and sprinkle with the two crumbled stock cubes as the onion softens. After two minutes, add the liquid stock, increase the volume with water and stir in some (whisper it quietly) gravy granules at the end to thicken the gravy!

If there's a lot of fat left with the meat juices after you've roasted the joint you can pour some of it away, but you'll need a little of it to get flavour into the onions.

I use pretty much the same method for any gravy made after roasting meat. For chicken, I'd omit the onions, but sometimes add different herbs, such as sage or perhaps some roasted garlic to the pan.

Ascension Bay Chilli

Serves 6

As a child, I loved David Attenborough's *Zoo Quest* books that I found on my father's bookshelf. They captured my imagination with dramatic tales of the capture of Komodo dragons, a black jaguar and a huge anaconda in distant jungles.

In 2024, I made a trip to Ascension Bay in Mexico on a fly fishing trip, in pursuit of the huge tarpon that navigate their way along this part of the Yucatan Peninsula. After two fruitless trips to Cuba over the past 30 years, I finally managed to catch and release 3 of those wonderful chrome-plated beasts, weighing up to 80lbs.

Lily, who runs the fishing lodge, gave me some dried chillies to take home, as I'd been telling her about my interest in cooking, and the cooking demonstrations that I do on social media. As well as Lily's chillies I returned home with a suitcase full of hot sauce, chocolate and lots of exciting cookery ideas.

I combined some of those Mexican ingredients in what is a tasty and authentic chilli that harnesses the flavours of one of the most interesting cuisines in the world. I decided to use butter beans because I like them more than red kidney beans. What better reason do you need?

This is another recipe that works really well in the Dutch oven that I use for so many of my slow-cooked dishes. If you're after something to warm you up on a frosty winter night, this fits the bill perfectly.

Ingredients

3 dried ancho chillies, stems removed and de-seeded
3 dried guajillo chillies, stems removed and de-seeded
1 litre of beef stock
Sea salt and black pepper
1.5kg of beef brisket, chopped into 2-3cm pieces
3 tbsp of olive oil
1 large onion, finely chopped
1 red pepper, diced
1 green pepper, diced
6 cloves of garlic, finely chopped
1 tbsp hot chilli powder
1 tbsp smoked paprika
1 tbsp of ground cumin
2 tsp dried oregano
2 tins of chopped tomatoes
2 tins of butter beans
1 tsp caster sugar
3 tbsp masa harina (you can use cornflour mixed with a
    little water to thicken the chilli if this is not available)
25g dark chocolate (min 70% cocoa)

Method

Preheat the oven to 150°C.

Dry roast the chillies in a pan on a medium high heat for 3 to 4 minutes until they start to smoke. Add enough water to cover the chillies, bring it to the boil then let it cool while the chilies rehydrate for 30 minutes.

Recipe continues overleaf

Drain the water from the rehydrated chillis and add to the beef stock, then blend with a stick blender.

Season the diced brisket generously with ½ tsp of sea salt and plenty of freshly ground black pepper. Add half the beef to a large cast iron casserole dish with 1 tbsp of olive oil and brown for 4-5 minutes, then place to one side and repeat with the rest of the beef.

Turn the heat down slightly and add the chopped onion and peppers, with another tbsp of olive oil, to the same pan and sauté for 2 minutes. Add the chopped garlic and cook for a further minute. Add the browned beef back to the pan and cook for 4-5 minutes.

Add all the spices and oregano to the beef in the pan, stir and cook for 1 minute until fragrant.

Stir in the liquidised chilli stock.

Add the chopped tomatoes and sugar.

Stir in the masa harina. Place the lid on the pan and cook in the oven for 2.5 hours or until meat is tender.

Add the beans and cook for a further 15 minutes.

When you're happy the chilli is cooked, stir in the chocolate until melted and serve with sides of rice, grated cheese, sour cream, salsa and a sprinkling of chopped spring onions.

Twice Baked
Jacket Potatoes

Serves 4

These are great as part of a hot buffet at a party, or as an accompaniment to a meal. However you choose to have them, they're delicious. They can be made ahead and even frozen, so you can celebrate your Christmas or New Year party with a minimum of fuss.

It's amazing how something as simple as the humble potato can become a rich and indulgent treat with the addition of some double cream, cheese and butter.

Ingredients

4 large baking potatoes
2 tbsp oil
4 slices smoked streaky bacon
4 spring onions, finely sliced
170g grated mature cheddar cheese
Large knob butter
2 tbsp double cream
Salt and pepper
Sour cream
Chives

Method

Preheat the oven to 200°C. Wash the potatoes and puncture the skin several times with a fork. Rub with oil, place on a baking tray and cook for one hour.

Allow the potatoes to cool somewhat, then slice in two and carefully scrape out the inside into a bowl.

Line up the empty skins on a baking tray.

Dice the smoked bacon, fry until crispy and add to the bowl. Add the spring onion, cheese, butter and cream. Season with salt and pepper. Mix until the butter has melted and the cream has incorporated into the mix.

Spoon the contents of the bowl back into the empty skins and bake for 30 minutes. Served topped with a dollop of sour cream and a sprinkling of chives. I guarantee they're delicious!

Note

This is an incredibly adaptable recipe. Why not use creme fraiche, or cream cheese instead of cream – try them with blue cheese if you like it. Finely dice vegetables such as onion, courgette or pepper and fry (why not use the bacon fat for extra flavour) before adding to the mix. Try different cheeses, particularly those that melt well, like Gruyère or Emmental. Add parmesan and pesto to give them an Italian twist. This really is a recipe you can experiment with forever.

At Christmastime, why not add finely chopped leftover turkey or other cooked meats to the mix, and add some chopped sage to give them a festive feel.

Honey Roast Ham

Serves 4

Ingredients

1.5kg ham, must have a good layer of fat
1 large onion, halved and studded with cloves
2 carrots
2 sticks of celery
3 or 4 bay leaves
12 black peppercorns
Handful of cloves
2 heaped tsp dark brown Muscovado sugar
70g clear honey
2 tsp dijon mustard
2 tsp wholegrain mustard

Method

Add the ham to a large pan of water that covers it, along with the onion, carrot, celery, bay leaves and peppercorns.

Put the ham on to boil for around 20 mins per 450g. Check the temperature with your probe, it needs to reach an internal temperature of 60°C. Combine the mustards, honey and sugar to make a glaze. Remove the cooked ham from the pan and cut off the top layer of skin with a knife, leaving the softer fat underneath.

Preheat the oven to 180°C. Score the fat on the ham in a criss-cross pattern of diamonds of about 2-3cm across and push a clove, into the centre of each diamond.

Place in a shallow roasting tin and pour the glaze over the ham, making sure the top is fully coated.

Cook in the oven for approximately 25 minutes, until the top is golden and sizzling.

Baste the ham in the juices from the bottom of the tin and cook for a further 5 minutes.

Remove from oven and allow to rest for 20 minutes before serving.

If you have leftovers, dice them up and use them in Farmer Cynthia's Turkey and Ham Pie (see page 172).

Farmer Cynthia's Christmas Leftover Turkey and Ham Pie

Serves 4

There are times when you like to make the effort, and there are times when you've already made the effort, and you'd appreciate a slightly easier life. Sometimes, when I'm making this pie, I stud an onion with cloves, add carrot, celery, bay leaves and perhaps a bouquet garni to the milk. On other occasions, and particularly after a belly bursting Cannon Hall Farm Christmas dinner, a simpler route just has to be taken. I've chosen to use ready rolled pastry for this pie, because after a busy Christmas no one needs the hassle of making pastry. This is a simple, tasty and wholesome way to use up those Christmas leftovers.

Ingredients

600g cooked turkey, diced into bitesize pieces
350g cooked roast ham, diced into bitesize pieces
100g butter
1 large leek, sliced thinly
75g plain flour
750ml whole milk
Salt and pepper
1 sheet of ready-rolled shortcrust pastry
2 eggs, beaten

Method

Preheat the oven to 200°C.

Place the cooked meat in the pie dish and spread it out evenly. Add 25g of the butter to a frying pan and sauté the leek on a low heat until soft. Spread the cooked leek evenly across the meat.

In a large saucepan, make a roux by melting the remaining butter and stirring in the plain flour until blended into a smooth paste. This should take 3 minutes or so.

Start to add the milk, little by little, whisking as you go to avoid lumps. As the mixture becomes smooth, add more milk bit by bit. The bechamel sauce should be of a medium-thick consistency. Add a little extra milk if you think it's too thick. Season with sea salt and black pepper.

Pour the bechamel into the pie dish, covering the meat and leeks evenly. Drape the rolled shortcrust pastry over a rolling pin and lay it evenly across the pie dish, crimping the edges so the pie is sealed. You could get creative with the decoration using any offcuts.

Make a small cross in the centre of the pastry so air can escape while cooking and brush with beaten egg.

Bake the pie for approximately 25 minutes, checking it after 20 minutes to ensure the pastry isn't burning. If you feel it is starting to burn, cover with foil for the last few minutes. Push a metal skewer into the centre of the pie and carefully touch to feel that it is hot all the way through.

Serve with vegetables of your choice.

Note

This recipe is for a pie dish that holds approximately 2 litres.

The Farm Shop

Dad had been farming at Cannon Hall Farm since 1958 and the prices he received when taking animals to market had always been disappointing. As we moved towards the end of the 1990s, it became obvious that there was no money to be made from selling animals at the local market. A fiver for a lamb seemed like scant reward for a lot of hard work.

I remember going to Barnsley Market with him when I was just a young lad. The auctioneer, Ken Green, was a giant walrus of a man to my child's eyes. Although there were no tusks present, his face hosted a monstrous moustache that had serious gravitas! I could easily imagine him scoffing oysters by the dozen like the walrus in Disney's *Alice in Wonderland*! He'd be there, doing his thing, as only he could, selling animals hard and fast, microphone pressed tight to his lips, spit flying in all directions, the stick that served for a gavel coming down hard against the battered desk in front of him followed by a deep, brusque 'Sold, Birkinshaw!'

I loved that place, the hustle and bustle around the selling ring, the buzz of conversation. The heady smell of the animals permeating the air and mingling with the beery aromas drifting from the bar. I remember the queue of cattle wagons that you walked past, drivers washing them out with hosepipes ready for the next load. Steam rising in the chilly air from both the animals, and from the warm Cornish pasties, clasped firmly in our little hands, that Dad had bought us from the market cafe. Mince, carrots and potato, a peppery delight in crumbly pastry – as always food was amongst the best of the memories.

Being with my dad and my brothers. Being a part of it all.

For many local stockmen, visiting the market was pretty much the only time they left their farms. It was so much more than a place to sell your cattle, pigs and lambs. It was a chance to meet like-minded people and I'm sure it served as therapy for many of the farmers and butchers that visited it. Now it's gone, replaced by garages selling cars. Very few of those markets in little towns remain. No longer needed they say, no longer relevant. Meanwhile depression and suicide amongst farmers, working long hours, often isolated in their tractor cabs or working on desolate hills and fells, is on the rise. The market closed in 1993, other local markets struggled on for a while but eventually they all fell by the wayside.

The time had come to make another big decision. We decided there had to be another way.

We would create another social hub for the community. We'd cut out the middleman and sell direct to the public on the farm. The idea for a farm shop was born. We started off small, a household freezer in the gift shop selling frozen beef steaks and joints packed up in plastic bags with no ceremony, but it began to sell.

Dad's sister, my auntie Shirley, was closing her farm shop at around the same time and Alan Asquith, her butcher, was looking for a new job. Shirley was only up the road at Redbrook, and Dad wouldn't open a farm shop in competition until she decided that the time was right to close for good. At the same time, John Holmes was selling his butcher shop in Dronfield. It seemed, staff for the new shop were available. We have huge gratitude to these two – both excellent – butchers who became firm friends and put us on the path to becoming one of the best farm shops in the UK. They taught us the basics, and Robert and I spent many hours with them in the butchery, learning what it took to turn out a great product.

The Farm Shop opened its doors for the first time in 1998. We started modestly, with one small butcher's counter, and a second-hand, walk-in meat fridge we got from Uncle Ted who was closing his butcher's shop.

Things started pretty slowly. John put out a very attractive display and showed us how to make some delicious sausages. In the beginning it was just farm customers having a look around, passing trade, really, with very few people coming just for the butchery. John was worried. It was good stuff! He couldn't understand why we weren't selling more! He wasn't sure if trade was ever going to pick up. Once again, we had to be brave. We'd built it and they would, we had to hope, eventually come!

We kept doing the right things, great quality produce at the right price and slowly word spread. Customers began to come through the door more and more regularly.

After a year or so, trade was growing quickly and we added another butchery counter. In the next few years, we added a separate shop with a deli counter and then, a few years later we built another extension to join the farm shop and deli together. A few years later we extended still further. It was the way things seem to happen. Our growth, as always, was through evolution rather than revolution.

We now have a butcher's counter full of home-produced beef, pork and lamb. We produce around 20 varieties of sausages on site and it's a thriving enterprise. Dad has come a long way from struggling to get five pounds for a lamb, but I'm sure he would love to walk back into Barnsley market and chew the fat with some of his old pals there.

In his 80s, the auctioneer, Ken Green, who I'd found such a memorable presence when I was a child, became a regular in our farm shop. Sadly, he's no longer with us, but it was great to hear from his son how much he loved our farm shop pork pies.

Farmer Roger's Roasted Partridge with Bread Sauce

Serves 2

Game is great! It's lean, healthy meat that's full of flavour. It's lived a good life, ranging freely in the countryside. It goes well with other strong flavours like foraged hedgerow fruit, mushrooms and woody herbs. The things you find in nature living alongside game generally go well with it.

One of my dad's favourite game birds is the partridge. Sadly, the English partridge is in decline but there are lots of French partridge readily available between 1st September and 1st February each year if you know where to get them. We keep a good stock of fresh game birds in the farm shop during the shooting season and often have frozen ones available later in the year.

Ingredients

2 oven-ready partridge
4 rashers streaky bacon
Sea salt and black pepper

Method

Dad likes his food plain and simple, and there's absolutely nothing wrong with an uncomplicated approach.

Preheat the oven to 225°C.

Season the partridge with salt and pepper, wrap them in the bacon slices and place in the oven on a shallow roasting tray.

The cooking time will vary with the size and maturity of the bird. Roast for 20-30 minutes, checking the temperature after 20 minutes with your probe. You're looking for a temperature of 65°C in the thickest part of the breast.

Rest for 5 minutes before serving.

Bread Sauce

Dad loves bread sauce above and beyond pretty much anything else and has his own way of making it. Getting a recipe out of him is very difficult, not because he's evasive, but because he's never written anything down and the answers to questions are almost unfathomable: 'you know, just the right amount,' or 'you know, just until it's done!'. I think I managed to work it out. This is a must at every Nicholson Christmas and is great with turkey, chicken or game birds. Dad would eat it with almost anything.

Ingredients

1 large onion, thinly sliced
6 cloves
200ml water
500ml whole milk
150g white breadcrumbs

Method

Add the onion and cloves to a medium size saucepan with enough water to cover the onion. Bring to the boil and let it simmer gently for approximately 20 mins, until soft. Add a little more water if necessary as you don't want the onion to brown or burn. You will need to watch the pan and stir occasionally.

Add milk and stir. Simmer for 10-15 minutes.

Add the breadcrumbs, heat gently for 4 or 5 minutes. Serve with potatoes roasted in goose fat.

Diced Potatoes Roasted in Goose Fat

Ingredients

3 tbsp goose fat
450g floury potatoes, diced into 1in cubes
Sea salt

Method

Preheat the oven to 200°C.

Spoon the goose fat into a heavy gauge baking tray and put it into the oven while you start the potatoes.

Boil the potatoes for 10 minutes and drain them using a colander. Shake them gently to roughen up the edges. This helps to give the potatoes a crispy outer layer. Add them to the hot goose fat and turn them to make sure they are evenly coated.

Give them a sprinkling of sea salt and roast for 60 minutes, turning them halfway through the cooking time, until they're brown and crispy on the outside and fluffy inside.

Happy New Year

One of the huge sadnesses of the New Year is that we now celebrate it without many of the characters who made it special in years gone by.

We'll never forget John Beever and Philip Armitage playing everyone's favourite party game! It was a simple affair, that involved holding a column of 2p pieces, clenched firmly between your tensed buttocks, shuffling forward until you were standing over a durable glass or small bowl, skilfully relaxing the said buttocks, and unleashing your small change into the receptacle below. The winner was the one with the most coins in the glass, and it would invariably cause huge guffaws of laughter from everyone watching. You would not believe just how competitive two rosy-cheeked old farmers can be! Even after John's hip replacement he'd still be first to volunteer – in fact he would insist on this game being played every single year! He'd hobble forward, his face a picture of concentration, bum cheeks firmly clenched, before depositing his load expertly in the glass. Such was the accuracy of his aim we were often left wondering if he'd been practicing at home!

I dare you to try it! It could become a family favourite at your house too. Do avoid using the best glasses though…

John and Philip are no longer with us, but the legend of their perfectly controlled buttocks lives on in perpetuity.

What Dad Brought
Back From The Zoo

After the farm had become established, we bought in a number of exotic animals in the years before a zoo licence became a requirement. These included wallabies, maras, rheas and emus.

After a trip to Twycross Zoo, Dad arrived home with some Cameroon Sheep and a Dwarf Zebu cow. It was a small, innocuous looking animal, not much larger than a sheep. It had a humped back and smallish horns, and it seemed to be a fairly docile creature.

One of my jobs, in between all the other farm jobs that needed doing, was part-time photographer. One day, I was in the cow's pen taking photographs as an old lady looked on from the viewing gallery. I was getting some good shots, head on, using my flash and wasn't feeling particularly threatened. Unbeknown to me, Jim, our butcher, was watching from above through the farm shop window that offered a great view of the stockyard. He was quietly linking sausages when he remarked to one of the other butchers, 'if he doesn't get out of there right now, that's going to have him.' As well as being a butcher, Jim was an experienced stockman who kept cattle... and he was right.

I retreated slowly into the corner of the barn, taking a photograph every few steps. The Zebu was advancing, tossing its head restlessly on its strong neck. I should have known, really... it all happened very suddenly! Before I knew what was going on, I was on my back in the corner of the pen, wrestling with the horns of a very angry dilemma! Three hundred pounds of angry zebu! The cow's horns were smashing into my legs. My camera was precious, and I wondered how I was going to get out of this keeping it in one piece. I struggled to my feet, in pain, trying to hold the grumpy Zebu by one horn, with my left hand holding out my camera bag to the old lady on the other side of the fence. She took it from me, followed by my camera.

Although I was still taking some fearsome punishment, I managed to get hold of both its horns, but it was powerful, and I had to get out as quickly as I could before it did any more damage. I managed to shove it back and made my escape, climbing out of the pen as fast as possible as the Zebu's horns crashed into the concrete wall where I'd just been. Those horns had left my legs black and blue, and I was suffering from a fair degree of bruised pride, too.

I thanked the old lady for her help and limped off to nurse my wounds and battered ego.

On her way out of the farm, the old lady was heard to remark 'was the young man that got gored by the bull alright?'

That cow hated me until the day it died. Whenever I walked past, it's eyes would follow me. I never went in the pen with it again. I knew it would have loved to have another go! As far as I was concerned, the best place for that cow was in Farmer Richard's Shin of Beef Stew with Red Wine recipe.

Farmer Richard's Shin of Beef Stew with Red Wine

Serves 4

Everyone loves a good stew, and for me this represents the perfect comfort food on a winter evening. I prefer to cook my shin on the bone. The bone marrow adds flavour and a lovely gelatinous quality to the finished dish. The connective tissues in the shin will soften and melt as you cook it slowly over several hours.

For me, shin is a fantastic choice, being great value for money, beautifully tender, but still holds together after the cooking process. I never cook with a wine that I wouldn't drink and this recipe gives you half a bottle left over to sip while you're waiting for the stew to cook. There are lots of reasonable value reds around, and I usually use a Côtes du Rhône when cooking this dish. I love my cast iron Dutch oven for this, but any decent heavy casserole dish with a lid will work fine. I start the cooking process on the top of the stove, then move it into the oven.

Ingredients

Plain flour
Salt and black pepper
Two or three good glugs of olive oil
4 generous slices of beef shin at least 1in thick, on the bone
1 large onion, diced
2 sticks of celery, diced
2 carrots, diced
2 bay leaves
Small bunch of thyme
Small bunch of rosemary
1 whole head of garlic, sliced through the middle, not peeled
3 beef stock cubes
Half a bottle of red wine
Water or beef stock
1 tsp Worcestershire sauce
12 shallots, peeled but left whole
Cornflour (or beef gravy granules) *optional*

Method

Preheat the oven to 150°C.

Place the flour in a bowl and season with salt and black pepper, then dust the pieces of beef in the seasoned flour.

Heat the oil to a medium/high temperature in the casserole and brown the meat on all sides. Set them aside on a plate.

Use a little more oil if needed and add the onion, celery, carrot, bay leaves, thyme, rosemary and garlic to the pan. Fry on a low/medium heat for about 10 minutes, until the vegetables are softened and the onion is translucent.

Crumble in the stock cubes, stir and then add the wine, deglazing the casserole dish. Increase the heat to medium and reduce the quantity by half. Add the beef back into the pan and cover with water, or beef stock if you want a richer flavour.

Add the Worcestershire sauce, put on the casserole lid and transfer to the oven for 4 hours.

Add the shallots and cook for a further 45 minutes. Sometimes I'll add some sliced carrots or small whole mushrooms half an hour from the end of the cooking time. If you feel your stew needs thickening, stir in a tablespoon or two of cornflour mixed with a little water. Gravy granules will do the job too! (It feels like a sin to mention them in a cookbook, but don't knock it if it works!)

Serve in a generously proportioned bowl with Roasted Garlic Mashed Potato (see page 190).

Note

I usually don't really measure anything when I'm cooking this, I just bang it all in there at the appropriate time and it seems to become delicious as if by magic!

Roasted Garlic Mashed Potato

Serves 4

Ingredients:

1 head of garlic, bottom sliced off so cloves are exposed
Sea salt
Black pepper
Olive oil
1kg of floury potatoes, peeled and cut into 1½in pieces
A generous knob of butter
A little milk

Method

Sprinkle the cut part of the garlic with salt and black pepper, drizzle with a little olive oil and wrap in a little parcel of foil. Roast in the oven at 180°C for 40 minutes, until soft and brown.

Boil the potatoes for around 20 minutes. When a sharp knife slips into the potato easily, they are done. Drain and leave to stand for a couple of minutes.

Mash the potatoes well. A potato ricer is brilliant for producing super smooth mashed potato; one pass through the ricer will do the job.

Squeeze out the softened garlic from the roasted head and add to the mashed potatoes in a bowl, along with the butter, a generous pinch of salt, several grinds of black pepper and a splash of milk. Beat the mixture together. I like to use a wooden spatula for this.

Taste and add more seasoning if required, or a little extra butter and milk if the mash is too stiff.

Rosemary's Meringues with Macerated Raspberries

Makes approx. 6 - 8 good size meringues

These meringues were one of the most popular offerings in Home Farm Tearooms when it opened back in 1981. Mum's friend, Rosemary, has been a constant in all our lives and a culinary inspiration. We've known her since we were small children, and she is always interested and engaged in whatever is going on at the farm. She's a larger-than-life character; a great cook and connoisseur of good food and she's been a tremendous support and inspiration in my food journey. I just couldn't produce a Cannon Hall Farm cookbook without trying to recreate the magnificent meringues that she is so expert at baking! They have always been served very simply with just a huge dollop of whipped cream, but here I've paired them with macerated raspberries, combining the sweetness of balsamic vinegar with the sharpness of the berries.

Ingredients
4 egg whites
225g caster sugar

Method
Preheat the oven to 100°C.

Prepare 2 baking trays/sheets with baking parchment. Spraying the baking sheet with a cake release spray before pressing on the baking parchment stops it slipping off the tray.

Whisk the egg whites until light, stiff and fluffy and you're able to turn the bowl upside down without the mixture falling out.

Continue whisking, adding a tablespoon of sugar at a time, until half the sugar has been whisked in. Add the rest of the sugar to the bowl and fold in carefully so you keep as much air in the mixture as possible.

Use 2 large spoons to shape the mixture into meringues and place them on the baking sheets with some space in between. Rosemary says you should bake them in the oven for 2 hours. We added another 20 minutes to Rosemary's timings. How long they take to cook can vary with the size of the meringues and the efficiency of the oven. Conventional wisdom says they should be dry all the way through, but we always preferred them slightly chewy in the middle.

Once cooked, turn the oven off and leave the meringues to cool slowly with the oven door ajar.

Ingredients

Macerated raspberries

220g raspberries
2 tbsp icing sugar
3½ tbsp balsamic vinegar

Method

Gently mix the raspberries, icing sugar and balsamic vinegar together.

Leave to stand for 30 minutes at room temperature before using.

Note

Everyone's oven is different, and nowhere is this more true than when making meringues! Ring the changes and adjust the temperature and timing to suit your oven. You may need a couple of attempts to get the meringues completely as you like them!

A Toast To Old Farmers

The old farmers of the nation have been under a lot of pressure lately. Government changes to inheritance tax, and unpredictable weather related to climate change is making growing the nation's food ever more challenging. Carrying on in the industry with a family farm is becoming less and less appealing. It is a vocation for many as much as a career. Now farming, like every other industry, has become more industrialised and wrapped up in red tape.

Government really should be encouraging small family farms to continue, not handing all the advantages and tax breaks to huge industrial farms owned by faceless corporations. The public want to know where their food comes from, and here at Cannon Hall Farm we are committed to continuing the traditional approach. We are determined to give them the option to buy traditionally reared beef, pork and lamb, direct from the farm, both online at cannonhallfarm.co.uk and in our farm shop.

And – as a well-deserved antidote to the trials and tribulations of modern day farming – many an old farmer would recommend a warming slug of sloe gin on a cold winter's day!

Sloe Gin

It's a simple process to produce sloe gin, but like many of the best things in life, more than a little patience is required.

The fruit of the blackthorn bush, fat black sloe berries, are a common and welcome sight in our hedgerows in the autumn. Countryside wisdom would dictate that they are at their best a little later in the season, after the first frost has wrapped its icy fingers around their fruiting bodies. If your sloes are collected early in the season and are firm to the touch, it's helpful to prick their skins to help release their flavour into the gin.

Steep the sloes in the gin in an airtight container for three months or so. The last lot I made had been steeping for around 18 months and it was epic, so don't worry if it's hanging around a little longer than you originally planned.

Ingredients

300g sugar
70cl gin
500g sloe berries

Method

Add the sloes to the gin (see facing page), and store for around 3 months. Then strain through a muslin cloth when you're ready to make the sloe gin.

Make up a sugar syrup by mixing the sugar with 150ml water. Bring it to the boil, stirring until all the sugar has dissolved, then allow to cool.

Add half of the sugar syrup to the gin and taste.

Continue to add and taste until you are happy with the balance between the tartness of the sloes, and the sweetness of the sugar syrup. They truly are a match made in heaven! Decant into a favourite bottle and enjoy.

Note

In the depths of winter, a warming glass of sloe gin is the perfect libation to enjoy, curled up in my favourite armchair, in front of the gently flickering fire, watching the snow fall steadily through the stone mullioned windows. It's a scene that, thankfully, has changed little over the centuries that this humble cottage has stood here. I imagine many an old miller has sat in this very room, gently illuminated by the firelight, their clothes still dusted in flour from the day's work. Ghosts of thoughts from long ago hang in the still air as I take a sip from my glass.

One Final Recipe...
Scabby Mushroom and Manky Carrot Soup

This dish is a homage to Mum and Dad's constant quest to avoid food waste. Having grown up in times of austerity when rationing was a huge part of everyday life, they quickly learned to appreciate any food that came their way. There is so much food wasted in modern society, and I can always spot my dad's nose wrinkling in disgust when 'good food' gets thrown away.

I recently asked Mum what she was having for tea (up in the North of England tea actually means dinner). She replied, with a chuckle, 'Your dad is making scabby mushroom and manky carrot soup.' This immediately tickled my sense of humour.

Basically, they clean out the fridge, identify any vaguely edible vegetables that have been left over from other recipes, cut off pieces of mould, peel off the wrinkly skins, drizzle them with a little oil and roast them in the oven to develop the flavour. They put them in a pot, add chicken stock, salt and pepper, simmer for half an hour then liquidise. If they want to add a little luxury, they'll stir in some cream and warm it through. Mum says it usually tastes great, makes them feel good about the lack of food waste and it's a great way to explore unlikely food combinations!

They were keen to emphasise that you can make great soup with food that would usually find its way into the bin.

You cannot fail to admire that sort of attitude. The world wouldn't be in the mess it is now if everyone shared their philosophy about food waste.

The Price Of Everything And The Value Of Nothing

In the modern world we seem to value all the wrong things. The most precious memories we make with our children very rarely involve money.

Children value things differently and we as adults should embrace this. Seemingly mundane items can have hidden value. Children can make something with no value seem priceless. Walking along a beach with a child, searching for seashells can suddenly elevate the value of a humble cockle or whelk shell. Have you ever searched for conkers with a small child, and been unable to conceal your delight when finding a particularly large, shiny example? Something that has no real worth at all.

Many of us envy others with designer watches and expensive clothes and cars in the materialistic world we live in. Perhaps we should reflect on the fact that value is subjective and that little children might just have the world sorted out before they become slaves to screens and modern consumerism. The little things can have value too.

This emotional inflation is such a lovely thing, and something that we as adults can buy into. It certainly extends into the world of food. Nature is full of ingredients you just can't buy. As a forager there's beauty in finding the first patch of wild garlic of the year, that you know will lead to some delicious, pickled flower buds, later in the spring. The joy of spotting fat, black sloes on a blackthorn bush in the autumn, and with it the knowledge that in a few months you'll be able to enjoy the sweet heady taste of sloe gin, straight from the hedgerow. You can even find Penny Bun mushrooms in the UK, it's our name for Porcini, a truly wonderful seasonal ingredient. As a fisherman, I get joy in the first shimmering string of mackerel coming aboard in the summer, because I know it's going to mean a feast later. Nature's harvest is a wonderful thing, we simply need to recognise and appreciate it when it comes along.

Afterword

On Writing

I always thought I had a book of some sort in me, and that it would be something that I'd find rewarding to do. In my teenage years I probably imagined it would be a novel, perhaps depicting some dystopian nightmare world of the future. Now, a few decades later, a cookbook seems a much more constructive and uplifting choice.

Cooking really has opened my eyes to a different sort of creativity.

If you're reading these last pages, I'd like to express my gratitude that you chose to come on the journey with me. Thank you for reading the book and helping to make me feel like a relevant part of the Cannon Hall Farm story once more.

It's tough when you realise that you're lacking a goal, a mission and a meaning. Equally, it's uplifting when you realise you have people on your side, that want you to succeed and achieve.

I came to a realisation that the way to feel relevant again was through the things that I love to do; the activities that have made me feel more whole. For me, that has always been through fishing, through art, through cooking and foraging and spending time feeling at one with nature. Through that, you can find a way to reconnect with the people, places and feelings that matter to you. You just need to find the spark and from a spark you can kindle a flame. Without a direction, and a little self-discipline, sparks are easily extinguished. This book started as a spark of inspiration on a long plane journey. By the time that flight was over I'd written several thousand words. A start, and enough to inspire me to continue. Although we are years on from that now, I have finally achieved what I set out to do and become a published author.

Everyone has something that makes them feel better about life and often the answer is to give yourself the time to rediscover what you've loved in the past. It's easy to tell yourself that you'll start tomorrow, but the best time to begin is always today.

In essence this book is a self-help project, and I hope it will inspire other self-help projects. It doesn't have to be something that will become a published work, it can be as simple as sitting and writing down the family recipes that you love and that you'd like to see passed on to the next generation. That's what we believe in at Cannon Hall Farm, respect for those who came before us and hope and optimism for those who follow. We'd like this little parcel of land of ours to continue to provide for the generations to come, and hope that the books we've written will be proudly displayed on the bookshelves of those who follow. If Marshall, Nelly, Arthur, James or their children or grandchildren are struggling, I hope they'll pick up one of those dusty old books from the shelf and look for a spark of light that will inspire them, and show them there is always a way forward, even when the dark might appear impenetrable.

Everyone has something special and unique in them. We all have the potential to achieve and succeed in something. It has been handed down in our DNA through the unique characteristics of our ancestors. We're all here for a reason and we can all make a difference. It's our responsibility to ourselves to live the best life we can. If you feel like you've lost your way, then don't look to others to put things right. The answer, and the catalyst for change is never to be found in some far-flung destination, it always lies within you.

My catalyst for change was this book. So, gentle reader, from the bottom of my heart, thank you for taking my hand and walking along the road with me.

Acknowledgements

I'd like to thank Jo Sollis, Clare Fitzsimons, Fergus McKenna, Claire Brown, Danny Lyle and the team at Mirror Books for believing in this, my first cookbook, and producing it to such an amazingly high standard. Thanks, too, to Amanda Stocks at Exclusive Press and Publicity.

Thank you to Clive Shalice and Victoria Gray for doing such a wonderful job with the photography and capturing the beauty and essence of both the food, and the lovely countryside around Mill Farm.

I'd like to thank my family, particularly Mum and Dad. Had it not been for their lifelong commitment there is no doubt that we wouldn't have a family farm. Without them, everything that followed on from opening the farm to the public in 1989, including this cookbook, would never have happened. I'd like to thank Robert and Julie, David and Anita, Tom and Simon, Katie and Phillip, and Poppy and Henry for their support. I hope my great niece Nelly and great nephews James, Arthur and Harold read and enjoy this book too – and learn to cook when they're older.

A special thank you to my much-loved son, Marshall, for his company, his humour and for always keeping me grounded and on my toes.

I'd like to thank my office buddies Helen, Mike, Amy, Michael, John, Jenny, Carol and Jo who've encouraged me along the way, tested recipes and proofread the text. Thank you to the hundreds of awesome people who work at Cannon Hall Farm, they are too numerous to mention individually, but they keep our well-oiled machine running every day and are integral to our success.

I'd like to thank all the followers and supporters of Cannon Hall Farm, many of whom have become friends. They have been so influential in supporting our family and promoting the farm, farm shop and this cookbook both through word of mouth and on social media.

I want to express my gratitude to all the many friends who have supported and believed in me over the years, most of whom are probably quietly astounded that I've managed to get this cookbook finished. To John and Sarah, Lee and Lisa, Grenville and Lorraine, Anthony and Sam and their lovely kids Alex, Beth and Joe. To the ever-entertaining Cleggy and Suzanne, to Rick and Zoe, Pete and Kim, Gemma and Paul, and Karen and David. Heartfelt thanks to Vicky and Kevin, Mike and Judith and Lee and Lynda who were Clare's friends initially, but who I now number amongst mine too.

Thanks to my late wife Maxine's family, including Yvonne, Errol, Marie and Lorenzo, Dave and Jo, Mikey, Maxie and Marcus, Val, Marc, Patsi, Cassie and Rick, Ayrton, Fran and Abbas, Hayley and their families, who remain firm friends.

I'd like to thank Matt and Cat of Pottery West. Their eyes met over a farmhouse loaf when they were students working in our farm shop and they now produce fabulous pottery together, some lovely examples of which are pictured in the pages here.

Thanks to Uncle Ted and his wife Jean for their constant encouragement and his ever-insistent demand that I believe in myself!

Thank you to my cousins Jayne and Anne who have encouraged, supported and given feedback throughout the process.

Thank you to Rosemary and Nigel and Margaret and Ron for a lifetime of friendship and support. You are wonderful people.

Thanks too, to my great friends Lizzie and Mike and their lovely children Jessica, Lucy and Katie for their friendship, support and the many great times we've enjoyed together over the years.

Thank you to my friends Carl, his wife Anne and their children William and Matias for their much-valued friendship and support.

Thanks to my old school pals, great friends and drinking companions on many a night out, Tree (Richard Sykes) and Did (David Pittaway) from my art college days in Barnsley. A special mention for Rob Newton, for many fun times and much sage advice. Last, but certainly not least, thank you to my great friend and fishing partner Matt Harris and his lovely wife Cath!

Finally, I'd like to thank my partner Clare, who has been a constant and guiding light, so influential and supportive in the journey towards getting this cookbook completed.

Index

Richard Nicholson, the oldest of the three brothers at Cannon Hall Farm, was born in 1966.

Instead of farming, Richard broke from tradition and received a degree in graphic design after studying in Hull.

In his spare time he enjoys watching sport, cooking, photography and fishing.

He now mostly works in the marketing department promoting the farm to the public and enjoys the varied aspects of social media.

He is also involved with buying for the gift and farm shops. He has a partner, Clare, and a son, Marshall.